Information Systems Engineering and Management

Volume 51

The book series "Information Systems Engineering and Management" (ISEM) publishes innovative and original works in the various areas of planning, development, implementation, and management of information systems and technologies by enterprises, citizens, and society for the improvement of the socio-economic environment.

The series is multidisciplinary, focusing on technological, organizational, and social domains of information systems engineering and management. Manuscripts published in this book series focus on relevant problems and research in the planning, analysis, design, implementation, exploration, and management of all types of information systems and technologies. The series contains monographs, lecture notes, edited volumes, pedagogical and technical books, as well as proceedings volumes.

Some topics/keywords to be considered in the ISEM book series are, but not limited to: Information Systems Planning; Information Systems Development; Exploration of Information Systems; Management of Information Systems; Blockchain Technology; Cloud Computing; Artificial Intelligence (AI) and Machine Learning; Big Data Analytics; Multimedia Systems; Computer Networks, Mobility and Pervasive Systems; IT Security, Ethics and Privacy; Cybersecurity; Digital Platforms and Services; Requirements Engineering; Software Engineering; Process and Knowledge Engineering; Security and Privacy Engineering, Autonomous Robotics; Human-Computer Interaction; Marketing and Information; Tourism and Information; Finance and Value; Decisions and Risk; Innovation and Projects; Strategy and People.

Indexed by Google Scholar. All books published in the series are submitted for consideration in the Web of Science.

For book or proceedings proposals please contact Alvaro Rocha (amrrocha@gmail.com).

Nazhatul Hafizah Kamarudin ·
Yusnani Mohd Yussoff · Tutut Herawan
Editors

Beyond Passwords: Protecting Healthcare with Authentication Security

Springer

Editors
Nazhatul Hafizah Kamarudin
Center for Cyber Security, Faculty of Information Science and Technology
Universiti Kebangsaan Malaysia
Bangi, Malaysia

Yusnani Mohd Yussoff
Department of Computer Engineering, School of Electrical Engineering
Universiti Teknologi MARA (UiTM)
Shah Alam, Malaysia

Tutut Herawan
Department of Information Systems, Faculty of Computer Science and Information Technology
University of Malaya
Kuala Lumpur, Malaysia

ISSN 3004-958X ISSN 3004-9598 (electronic)
Information Systems Engineering and Management
ISBN 978-3-031-92246-6 ISBN 978-3-031-92247-3 (eBook)
https://doi.org/10.1007/978-3-031-92247-3

This Springer imprint is published by the registered company Springer Nature Switzerland AG
The registered company address is: Gewerbestrasse 11, 6330 Cham, Switzerland

This book is dedicated to my dear parents, Kamarudin Abd Karim and Che Rahana Ab Rahaman. Your unwavering support has been my foundation. To my beloved husband, Khairul Mizan Mohd Fauzi, thanks for your constant presence and encouragement. Finally, to my precious children, Muhammad Al Ayyash and Sara Alyssa, this accomplishment is dedicated to both of you.

—Nazhatul Hafizah Kamarudin

This book is dedicated to the healthcare professionals, IT experts, and cybersecurity advocates who tirelessly work to safeguard our health systems. To my students, whose curiosity and enthusiasm inspire innovation, and to my family, whose unwavering support fuels my passion for this important work.

—Yusnani Mohd Yussoff

This book is dedicated to my beloved family.

—Tutut Herawan

Preface

Mobile e-health technologies are revolutionizing healthcare by seamlessly integrating the Internet of Things (IoT) to enhance medical performance and secure patient data. This book delves into the critical role of authentication in safeguarding e-health systems, advocating for stronger authentication schemes and the development of robust security protocols to protect sensitive healthcare information. Traditional healthcare systems, which rely heavily on centralized medical facilities, face heightened security and privacy risks in the e-health landscape due to the evolving threat landscape. The prevalent use of usernames and passwords in mobile e-health authentication exposes these systems to significant vulnerabilities, including replay attacks and sensor node cloning, which can compromise network security and patient data integrity. In response to these threats, this book introduces a secure, innovative non-regenerated unique identity-based authentication protocol designed specifically for wireless embedded e-health sensor nodes and mobile e-health applications. This approach ensures seamless and secure authentication, safeguarding sensitive healthcare data from unauthorized access. The book also explores the implementation and testing of mobile e-health applications, demonstrating how passwordless authentication can streamline communication between patients and healthcare providers while enhancing security.

Recognizing the reliance of many mobile e-health systems on third-party service providers, this book proposes a two-tier mobile e-health system framework that eliminates the need for third-party servers, thereby strengthening the confidentiality of network transmissions. A non-regenerated unique identity is generated for each e-health sensor node, forming the foundation of a mobile e-health authentication protocol that enhances the security of the entire network architecture. Additionally, a comprehensive security analysis using formal methods, including AVISPA, demonstrates the protocol's effectiveness in defending against replay attacks and sensor node cloning, ensuring a robust and secure mobile e-health environment. This book offers a forward-thinking solution to modern e-health security challenges, empowering healthcare systems to move beyond traditional password-based authentication and towards a safer, more efficient future.

Bangi, Malaysia
Shah Alam, Malaysia
Kuala Lumpur, Malaysia

Nazhatul Hafizah Kamarudin
Yusnani Mohd Yussoff
Tutut Herawan

Contents

Abbreviations

ABE	Attribute Based Encryption
AES	Advanced Encryption Standard
BAN	Body Area Network
BPM	Beats Per Minute
CA	Certificate Authority
DLP	Discrete Log Problem
ECC	Elliptic Curve Cryptography
ECG	Electrocardiogram
EDR	Enhanced Data Rate
GDP	Group Device Pairing
GUI	Graphic User Interface
HIPAA	Health, Insurance, and Accountability
HLPSL	High Level Protocol Specification Language
IBE	Identity Based Encryption
ICT	Information and Communication Technology
ID	Identity
IOT	Internet of Things
ISM	Industrial, Scientific, and Medical radio band
MAC	Message Authentication Code
NFC	Near Field Communication
PAN	Personal Area Network
PKG	Public Key Generator
PKI	Public Key Infrastructure
RBAC	Role Based Access Control
RC	Registration Centre
RFID	Radio Frequency Identification
RSA	Rivest-Shamir-Adleman Algorithm
SNEP	Sensor Network Encryption Protocol
SPAN	Security Protocol Animator
SSL	Secure Sockets Layer
TA	Trust Authority

UHM	Ubiquitous Health Monitoring
WBAN	Wireless Body Area Network
WHO	World Health Organization
WLAN	Wireless Local Area Network
WSB	Wireless Service Broker
WSN	Wireless Sensor Network

List of Figures

Authentication Strategies for Modern Healthcare Systems

Trusted Access: Authentication Solutions for Mobile E-Health

Protocols for Mobile E-Health Authentication

Performance Analysis of Mobile E-Health Authentication System

Strengthening E-Health Security: Authentication Analysis Through Formal Methods

List of Tables

Strengthening E-Health Security: Authentication Analysis Through Formal Methods

Exploring Authentication Scopes in Digital Healthcare

Nazhatul Hafizah Kamarudin, Yusnani Mohd Yussoff, and Tutut Herawan

Abstract The rapid adoption of Wireless Sensor Networks (WSN) and Internet of Things (IoT) technologies in healthcare has highlighted the need for secure, efficient, and user-friendly authentication protocols. Traditional password-based methods are increasingly inadequate, particularly for elderly users who may struggle with complex login procedures. This study presents the development of a secure, passwordless authentication protocol specifically designed for mobile e-health systems, addressing critical challenges related to identity verification, privacy, and data security. By leveraging a two-tier mobile e-health architecture that removes the dependency on third-party servers, this protocol enhances both system reliability and data confidentiality. The protocol uses non-regenerated unique identities of e-health sensor nodes, validated through a secure cryptographic boot process, effectively mitigating the risks of node cloning attacks and unauthorized access. Formal validation was conducted using HLPSL, AVISPA, and SPAN tools, demonstrating the protocol's robustness against replay and node cloning attacks. The approach also reduces the need for complex key management processes and is designed with energy efficiency in mind, making it well-suited for resource-constrained e-health environments. This authentication framework supports seamless access for users, particularly benefiting elderly populations, while simultaneously reinforcing security measures within the mobile e-health infrastructure. The findings contribute to advancing a more secure, practical, and accessible mobile e-health authentication system that meets the demands of modern healthcare applications and enhances user experience.

N. H. Kamarudin (✉)
Center for Cyber Security, Faculty of Information Science and Technology, Universiti Kebangsaan Malaysia, Bangi, Malaysia
e-mail: nazhatulhafizah@ukm.edu.my

Y. Mohd Yussoff
School of Electrical Engineering, College of Engineering, Universiti Teknologi MARA, Shah Alam, Malaysia

T. Herawan
Department of Information Systems, Faculty of Computer Science and Information Technology, University of Malaya, Kuala Lumpur, Malaysia

N. H. Kamarudin et al. (eds.), *Beyond Passwords: Protecting Healthcare with Authentication Security*, Information Systems Engineering and Management 51,
https://doi.org/10.1007/978-3-031-92247-3_1

Keywords Passwordless authentication · Mobile e-health authentication · Internet of things (IoT) · Healthcare data security

1 Introduction

E-health network can support integrated healthcare services and data interoperability to communicate in the form of electronic health records accessible through Wireless Sensor Networks (WSN) [1]. Even if the users are far from the healthcare centre, e-health monitoring devices can be applied to obtain the user data and to get access to the healthcare services [2, 3]. Since the last decades, WSN are continuously being utilized within the healthcare system in order to improve the efficiency and reliability of the healthcare operation system. Healthcare services costs are getting higher nowadays and therefore, it is essential for the healthcare organizations to consider adopting e-health system. It will allow the healthcare centre to modernize their health application processes and provide more efficient and reliable services in a cost-effective manner. With a fast growing development of the Internet of Things (IOT), e-health system provides a strong infrastructure and offer viable healthcare services over the network [4]. Utilizing the latest technologies in the healthcare services is an important approach for the healthcare organizations to enhance the healthcare services and reduce operational costs [5, 6]. High demand in healthcare services and shortages of qualified healthcare professionals show the need for e-health implementation by the healthcare organizations. New research advancements in WSN and Internet technology have assisted the development of e-health services [7].

The healthcare professionals also benefit from the development of the e-health system where they can save more time to attend patients and have the opportunity to attend more severe cases [8]. Besides, they are able to have a faster access to their patient health condition by integrating the e-health technology in the healthcare system [9]. The application of the IOT in healthcare system provides several advantages such as a secure patient identification and systematic data collection [3, 10]. Recent technology offers the ability to monitor patient condition autonomously through remote monitoring system [11]. By taking advantage from the IOT emerging paradigm, it is now achievable to monitor patient activity outside the healthcare centre. The growth of health issues nowadays has forced the healthcare organization to implement the e-health system in order to reduce the high cost by avoiding unnecessary hospitalization and giving priority to urgent cases [12]. Health monitoring platform using WSN will support various wearable sensor nodes and it has general computing ability through embedded system [13, 14].

Most existing e-health systems concentrate on the patient data acquisition and medical embedded component with little attention to the other important aspects such as real time monitoring and most importantly the security of the e-health system [15]. Wireless body area network (WBAN) is the key research area in e-health application in order to create an intelligent and secured e-health system. With increasing numbers of people visiting the healthcare centre, the need for remote healthcare monitoring system is highly needed to assist the situation. Therefore, advancement in electronics and biomedical applications have become a growing attention recently since these areas contribute in e-health monitoring system [15]. The cost of e-health monitoring system is no longer an issue since it is now affordable for everyone to have a mere smartphone to get connected with e-health. Wearable sensor node will be deployed on the patient body for continuous monitoring and it can range from micro sensors to complex and high-tech sensor nodes.

However, since e-health sensor node is deployed in the patient body area network (BAN), it is known to be more vulnerable to security attacks. Hence, it is necessary to implement security features on the sensor nodes in order to protect the e-health sensor from node cloning attack. E-health sensor node that is mostly left unattended by the healthcare professional and physically accessible to adversary can expose the healthcare information and data confidentiality of the patient. On that particular reason, this research intends to design a passwordless authentication using a secure non-regenerated unique identity of the e-health sensor node to provide a secure and reliable sensor node identity for the e-health authentication protocol.

2 Identifying Challenges in Authentication for E-Health Systems

A secure e-health authentication system is one of the potential approaches in order to provide high security for the whole e-health monitoring system [16]. It is to ensure data confidentiality and the authenticity of the e-health sensor node besides implementing a secure mobile e-health application for wide range of people especially the elders. Increasing numbers of elderly people with acute illnesses has led to a high demand for support services in healthcare system and it is important to define new model for e-health system in order to assist continuous care for patients mostly elderly people [17–19]. Since elderly and aged people are the target user in the healthcare organizations, e-health system design needs to be user-friendly and accessible for people of all generations.

Unlike conventional wireless sensor network (WSN), mobile e-health deals with medical information and has a stringent requirement for security and privacy. Since mobile e-health transmits sensitive and confidential data, a strong and secure authentication is required to be implemented in the system. However, most authentication schemes in e-health monitoring system are using username and password for identity verification [13–22]. Besides vulnerable to security attacks like password break and hacking, this mobile e-health authentication scheme is not suitable for elderly people. Elderly people are known to have difficulty of remembering the username and password combination [18].

Involvement of a third party server or e-health service provider in most e-health current architecture has caused lack of reliability in the system. The e-health services have been extended outside its own architecture since most of the e-health systems depend on a third party server to transmit and store the data [23, 24]. Plus, for some mobile e-health applications, the third party server also is needed to generate the security parameters for their authentication protocols [25–27]. Privacy and confidentiality of the data should be prioritized and therefore it is very important to eliminate the need for a third party server in the network architecture. Improving the e-health current three-tier architecture into a two-tier mobile e-health architecture will enhance the security of the whole e-health system.

Security challenges in mobile e-health system will lead to an e-health transformation and with the development in IOT applications, mobile e-health security systems need to be improved and upgraded [28]. Acknowledging the current security problems with the e-health sensor node and mobile device limited constraints facing by mobile e-health system, this study will present a comprehensive security method solution in term of security authentication protocol in order to provide a seamless and secure mobile e-health implementation while ensuring its practicality and efficiency.

The proposed e-health security authentication protocol is aimed to design a secure passwordless authentication for e-health monitoring system and at the same time validating only registered e-health sensor node in the network system. As energy consumption will increase linearly to the additional security levels of implementation, this proposed security authentication protocol will take into consideration the limited resources and constraints in the e-health networking environment.

The objectives of this book encompass three key aims. Firstly, to establish a robust framework for a secure mobile e-health authentication network architecture, achieved through the design of a two-tier e-health network capable of circumventing reliance on third-party servers or e-health service providers. Secondly, to develop a secure passwordless authentication protocol tailored for mobile e-health monitoring systems. This protocol will authenticate users by harnessing the non-regenerated unique identity of the e-health sensor node, ensuring a foundation of secured development. Lastly, the book aims to assess the feasibility of the proposed protocol, considering factors such as energy consumption, memory utilization, and comprehensive security analysis.

3 Contributions to Advancing a Secure Mobile E-Health Authentication Protocol

This book presents the following contributions towards the development of a secured mobile e-health authentication protocol. The identity of the e-health sensor node is a crucial element in this proposed mobile e-health authenticating process and by generating the non-regenerated unique identity of the e-health sensor node through a developed ideal secure boot cryptographic process, the possibility of sensor node cloning attack by the adversary can be eliminated. Username and password scheme is no longer needed since the system will validate the unique identity of the e-health sensor node generated through the secure boot process. Username and password scheme has been proven insecure since it can be cracked by the adversary. Hence, this mobile e-health system is safe from password break attack and identity theft hacking. The design of the developed secure boot cryptographic process has considered the power limitation and energy constraint faced by the e-health sensor nodes in the network. No extra security chip is implicated or attached with the e-health sensor node.

The developed passwordless mobile e-health authentication protocol contributes to a new method in authenticating the e-health sensor node as well as the user before getting into the mobile e-health system. In addition, demand for security key distribution mechanism is no longer needed due to the existence of security keys deployment in the registration phase. The design of the passwordless mobile e-health authentication protocol has contributed significantly towards the security establishment procedure during the mobile e-health authentication. Most importantly, the design of passwordless mobile e-health authentication protocol would provide a great assistance to people mostly elderly users since it is a seamless authentication without the requirement of remembering the username and password combination anymore. EHEART application for mobile e-health system is designed in order to facilitate the passwordless authentication protocol requirement and the power consumption of mobile e-health system can be reduced through its implementation.

To ensure that only trusted entities exist in the mobile e-health network, a framework of two-tier e-health architecture is developed. Two-tier e-health architecture provides a reliable networking environment for the mobile e-health system. By eliminating the third party server in the current mobile e-health architecture, data confidentiality and system reliability will be protected. User privacy is an important and critical issue in mobile e-health system thus making users too cautious regarding sending their information through the e-health network. By developing two-tier e-health architecture while maintaining the same function as three-tier e-health architecture, it has highly contributed to the cost reduction for the mobile e-health system. The e-health system will no longer depend on the third party server or e-health service provider for user authentication or data storage. The two-tier e-health architecture suggests e-health sensor node identity verification and eliminates the function for PKI such as e-health service provider as well as the need for certificate authority (CA). Moreover, this framework has contributed significantly towards a secure establishment process in the mobile e-health access control structure.

Passwordless mobile e-health authentication protocol is developed using High Level Protocol Specification Language (HLPSL). The protocol model is then formally validated through AVISPA tools and Security Protocol Animator (SPAN) in order to analyse the system security and practicality. The passwordless authentication protocol is validated using formal method and the mobile e-health system security is proven to be secured from replay attack and node cloning attack. The design of the mobile e-health authentication protocol has assisted in protecting the e-health sensor node and it can be demonstrated through the development of EHEART application prototype as well as through formal method analysis.

4 Conclusion

IoT has constructed a different prospect in WSN through e-health system where sensor node is not merely connected to a user, but currently it is connected to the whole surrounding system and is able to get access to the database server to store the confidential data for further analysis. It is leading the direction towards a better transition in e-health sensor networking. E-health can eliminate patient's limitation, enhance mobility, and reduce cost of healthcare services. Also, continuous health monitoring through e-health system can diminish the possibility of late health treatment. By integrating the healthcare system in WSN, it supports a lot of benefits for user with various disabilities due to the efficiency and accessibility of the e-health system. Advancements in mobile technology have allowed mobile devices, such as smartphones and tablets, to be used in a variety of different applications. With the connection of the sensor devices and the Internet, as well as the mobile phones have been widely used among people due to its ubiquitous feature, it is now becoming more feasible than before to use mobile technology for medical applications. A user can simply connect a wireless sensor on their body to a mobile device in order to monitor their health data. This chapter outlines significant contributions towards enhancing the security of mobile e-health authentication protocols. Firstly, it introduces a method for generating non-regenerated unique identities for e-health sensor nodes, enhancing security by thwarting cloning attacks without relying on vulnerable username-password schemes. The design, mindful of energy constraints, doesn't necessitate additional security chips. Secondly, a novel passwordless mobile e-health authentication protocol is proposed, streamlining access for users, especially the elderly, while obviating the need for complex key distribution mechanisms. This protocol, integrated into the EHEART application, reduces power consumption and enhances security through formal validation using HLPSL, AVISPA, and SPAN tools, safeguarding against replay and node cloning attacks. Additionally, a two-tier e-health architecture framework is introduced, enhancing trust and privacy by eliminating third-party dependencies and costly PKI services, thus contributing to a more secure and cost-effective mobile e-health system.

References

1. Hosseini, A., Emami, H., Sadat, Y., Paydar, S.: Integrated personal health record (PHR) security: requirements and mechanisms. BMC Med. Inform. Decis. Mak. **23**(1), 116 (2023). https://doi.org/10.1186/s12911-023-02225-0
2. Tan, Z.: A user anonymity preserving three-factor authentication scheme for telecare medicine information systems. J. Med. Syst. **38**(3) (2014). https://doi.org/10.1007/s10916-014-0016-2
3. Suraci, C., et al.: The next generation of ehealth: a multidisciplinary survey. IEEE Access **10**, 134623–134646 (2022). https://doi.org/10.1109/ACCESS.2022.3231446
4. Kamarudin, N.H., Suhaimi, N.H.S., Nor Rashid, F.A., Khalid, M.N.A., Mohd Ali, F.: Exploring authentication paradigms in the internet of things: a comprehensive scoping review. Symmetry (Basel) **16**(2), 171 (2024). https://doi.org/10.3390/sym16020171

5. Raj, A., Prakash, S.: A secure blockchain-based access control architecture for IoT-healthcare applications. Natl. Acad. Sci. Lett. (2024). https://doi.org/10.1007/s40009-023-01383-z
6. Jan, M.A., Zhang, W., Akbar, A., Song, H., Khan, R., Chelloug, S.A.: A hybrid mutual authentication approach for artificial intelligence of medical things. IEEE Internet Things J. **11**(1), 311–320 (2024). https://doi.org/10.1109/JIOT.2023.3317292
7. Aski, V.J., Dhaka, V.S., Kumar, S., Verma, S., Rawat, D.B.: Advances on networked ehealth information access and sharing: status, challenges and prospects. Comput. Netw. **204**, 108687 (2022). https://doi.org/10.1016/j.comnet.2021.108687
8. Chan, V., Ray, P., Parameswaran, N.: Mobile e-health monitoring: an agent-based approach. IET J. Telemed. e-Health Commun. Syst. **2**(2), 223–230 (2008). https://doi.org/10.1049/iet-com
9. Guo, Y., Hu, Y., Afzal, J., Bai, G.: Using P2P technology to achieve ehealth interoperability. In: International Conference on Service Systems and Service Management (ICSSM), pp. 1–5. IEEE, Tianjin (2011). https://doi.org/10.1109/ICSSSM.2011.5959430
10. Suleski, T., Ahmed, M., Yang, W., Wang, E.: A review of multi-factor authentication in the internet of healthcare things. Digit Health **9**, 20552076231177 1 (2023). https://doi.org/10.1177/20552076231177144
11. Ahmed, M.I., Kannan, G.: Secure and lightweight privacy preserving internet of things integration for remote patient monitoring. J. King Saud Univ. Comput. Inf. Sci. **34**(9), 6895–6908 (2022). https://doi.org/10.1016/j.jksuci.2021.07.016
12. Salvi, D., Mora, E.V., Teresa, M., Waldmeyer, A.: An architecture for secure e-health systems
13. Roozbeh Jafari, A., Encarnacao, A.Z., Foad Dabiri, S., Noshadi Majid, H.: Wireless sensor networks for health monitoring. In: International Conference on Mobile and Ubiquitous System: networking and Services, pp. 479–481. IEEE (2005). https://doi.org/10.1109/MOBIQUITOUS.2005.65
14. Thilakanathan, D., Chen, S., Nepal, S., Calvo, R., Alem, L.: A platform for secure monitoring and sharing of generic health data in the cloud. Elsevier Future Gener. Comput. Syst. **35**, 102–113 (2014). https://doi.org/10.1016/j.future.2013.09.011
15. Otoom, A.F., Abdallah, E.E., Kilani, Y., Kefaye, A., Ashour, M.: Effective diagnosis and monitoring of heart disease. Int. J. Softw. Eng. Appl. **9**(1), 143–156 (2015). https://doi.org/10.14257/ijseia.2015.9.1.12
16. Liu, C.H., Chung, Y.F.: Secure user authentication scheme for wireless healthcare sensor networks. Comput. Electr. Eng. **59**, 250–261 (2017). https://doi.org/10.1016/j.compeleceng.2016.01.002
17. Bazzani, M., Conzon, D., Scalera, A., Spirito, M.A., Trainito, C.I.: Enabling the IoT paradigm in e-health solutions through the VIRTUS middleware. In: 11th International Conference on Trust, Security and Privacy in Computing and Communications, pp. 1954–1959. IEEE (2012). https://doi.org/10.1109/TrustCom.2012.144
18. Bai, G., Guo, Y.: A general architecture for developing a sustainable elderly care e-health system. In: International Conference on Service Systems and Service Management (ICSSM), pp. 1–6. IEEE, Tianjin (2011). https://doi.org/10.1109/ICSSSM.2011.5959443
19. Archer, N., Fevrier-Thomas, U., Lokker, C., McKibbon, K.A., Straus, S.E.: Personal health records: a scoping review. J. Am. Med. Inform. Assoc. **18**(4), 515–524 (2011). https://doi.org/10.1136/amiajnl-2011-000105
20. Omar, W.M., Taleb-Bendiab, A.: E-health support services based on service-oriented architecture. IEEE J. Mag. IT Prof. **8**(2), 35–41 (2006). https://doi.org/10.1109/MITP.2006.32
21. Sudha, G., Ganesan, R.: Secure transmission medical data for pervasive healthcare system using android. In: International Conference on Communication and Signal Processing, Melmaruvathur, pp. 433–436 (2013)
22. Kamarudin, N.H., Yussoff, Y.M., Hashim, H.: IBE_TRUST authentication for e-health mobile monitoring system. In: IEEE Symposium on Computer Applications & Industrial Electronics (ISCAIE), pp. 160–164. IEEE, Langkawi (2015). https://doi.org/10.1109/ISCAIE.2015.7298348

23. Nagaty, K.A.: Mobile health care on a secured hybrid cloud. J. Sel. Areas Health Inform. (JSHI) **4**(2) (2014)
24. Srivastava, K., Awasthi, A.K., Kaul, S.D., Mittal, R.C.: A hash based mutual RFID tag authentication protocol in telecare medicine information system. J. Med. Syst. **39**(1) (2015). https://doi.org/10.1007/s10916-014-0153-7
25. Amin, R., Biswas, G.P.: A novel user authentication and key agreement protocol for accessing multi-medical server usable in TMIS. J. Med. Syst. **39**(3) (2015). https://doi.org/10.1007/s10916-015-0217-3
26. Lopes de Souza, R., Lung, L.C., Custodio, R.F.: Multi-factor authentication in key management systems. In: 12th IEEE International Conference on Trust, Security and Privacy in Computing and Communications, pp. 746–752. IEEE, Melbourne (2013). https://doi.org/10.1109/TrustCom.2013.90
27. Sood, S.K., Sarje, A.K., Singh, K.: A secure dynamic identity based authentication protocol for multi-server architecture. J. Netw. Comput. Appl. **34**(2), 609–618 (2011). https://doi.org/10.1016/j.jnca.2010.11.011
28. Saranya, A., Naresh, R., Karuppiah, S., Jenifer, M.: Development of trust-based authorization and authentication framework for secure electronic health payment in cloud environment. Soft Comput. (2024). https://doi.org/10.1007/s00500-023-09514-w

Authentication Strategies for Modern Healthcare Systems

Nazhatul Hafizah Kamarudin, Yusnani Mohd Yussoff, and Tutut Herawan

Abstract The study addresses the pressing need for secure, user-friendly authentication protocols in mobile e-health systems, with a focus on overcoming vulnerabilities in existing password-based methods. By leveraging a comprehensive analysis of mobile e-health architectures and current authentication strategies, this research identifies significant gaps in security, particularly in the context of Wireless Sensor Networks (WSN) and the Internet of Things (IoT) in healthcare. A two-tier mobile e-health network architecture is proposed, aimed at eliminating the dependency on third-party servers and reducing risks associated with node cloning and data breaches. This protocol incorporates a unique, non-regenerated identity for each e-health sensor node, validated through a robust cryptographic boot process. Formal security analysis conducted using AVISPA and SPAN tools demonstrates resilience against common threats such as replay attacks and impersonation. This approach not only reduces the complexity of managing cryptographic keys but also enhances energy efficiency, making it suitable for resource-constrained environments. The findings underscore the potential of a passwordless authentication framework to significantly enhance data security and simplify user interaction, particularly for elderly users, in mobile e-health applications. This framework sets the stage for a more secure and scalable mobile e-health infrastructure, offering improved protection of sensitive health data and a streamlined experience for both patients and healthcare providers.

Keywords Passwordless authentication · Wireless sensor networks (WSN) · Identity-based authentication · Authentication protocol · Mobile e-health

N. H. Kamarudin (✉)
Center for Cyber Security, Faculty of Information Science and Technology, Universiti Kebangsaan Malaysia, Bangi, Malaysia
e-mail: nazhatulhafizah@ukm.edu.my

Y. Mohd Yussoff
School of Electrical Engineering, College of Engineering, Universiti Teknologi MARA, Shah Alam, Malaysia

T. Herawan
Department of Information Systems, Faculty of Computer Science and Information Technology, University of Malaya, Kuala Lumpur, Malaysia

N. H. Kamarudin et al. (eds.), *Beyond Passwords: Protecting Healthcare with Authentication Security*, Information Systems Engineering and Management 51,
https://doi.org/10.1007/978-3-031-92247-3_2

1 Introduction

There has been significant research on the mobile e-health authentication system in order to enhance the efficacy and security of the e-health system. As demand for e-health application has increased, a secure authentication protocol has become the main concern in the e-health networking [1]. This chapter primarily introduces an overview of e-health system and existing mobile e-health applications followed by the analysis of security constraints in the mobile e-health networking. Then, mobile e-health network architecture is firstly critically analysed and discussed. Also, related works on existing mobile e-health authentication protocols as well as identity based authentication protocols are comprehensively scrutinized in order to discover the security issues in the e-health authentication system. Afterward, the assessment regarding the security requirements and security analysis in the mobile e-health authentication is also concisely discussed in this chapter.

2 Overview of E-Health

Improvement in communication technology as well as sensor networking has assisted the development of e-health application in order to provide effective healthcare services to the people. According to the World Health Organization (WHO), e-health is the transfer or delivery of healthcare resources and health information using the power of information and communication technologies (ICT) in order to improve current healthcare services [2]. With the rapid advancement in the area of sensor network, WBAN is then developed to control medical sensor nodes for e-health monitoring and to transmit the health data within a suitable communication network [3]. Wired connection in health monitoring system has become burdensome and no longer suitable since it can reduce patient mobility and flexibility. With the integration of WBAN, e-health network currently consists of medical sensor nodes and e-health platform application such as computers or mobile device communicating wirelessly with the healthcare base station or the e-health server. Other advantages of implementing e-health in the healthcare organization are well defined which include cost reduction [4–7], quality of care improvement [8–36], systematic data storage [37–58], and healthcare services efficiency [58–70]. Nonetheless, e-health system needs to achieve the consistency of security requirements. Implementing new technology in e-health mostly involves complex procedure and time consuming with only ambiguous outcomes. Over the years, a lot of research have been done focusing only on the e-health implementation process rather than analysing the outcomes and security requirements for the e-health system [8]. People mostly see e-health technology as a way to enhance the service efficiency and reduce errors however security and safety concerns in e-health has undermined the people confidence in e-health therefore weakens the e-health utilization in healthcare organization. Presently, lots of e-health sensor nodes have been developed using recent technologies to be low-cost, low power multi-functional sensor nodes that are also smaller in sizes in order to achieve an effective e-health monitoring system. However, the communication issues between the e-health sensor nodes and mobile device as well as with the e-health server are still enhanced since the system faces many security challenges in dealing with the e-health data transmission. It is required to provide a secure communication channel out of an insecure environment in the e-health system since it is transmitted over a wireless sensor network.

E-health basically consists of a huge scope of emerging technology which is part of the Internet of Things application that can be applied in the medical field. It is a scope of research between health informatics and health services that can be delivered and enhanced through information and communication technologies (ICT) [71]. However, e-health system as a whole does not necessarily provide the aspect of remoteness in the network. Telehealth on the other hand encompasses general concept of e-health such as educational researches on healthcare, health administrative uses, and medical applications through a distance in which the term 'tele' itself in Greek gives a meaning of distance. Furthermore, telemedicine provides more specific healthcare services that include information on patient diagnosis, healthcare treatment and research area on medical issues. With the advancement in mobile technology, a new capability of the e-health system is added to the e-health area that utilizes the function of mobile devices to acquire and transmit the e-health data. Mobile e-health or m-health system can be defined as the advance mobile applications and network technologies to enhance the healthcare services within a certain distance area. Mobile e-health has contributed to a great assistance in increasing patient mobility as well as facilitating healthcare professional's demands.

Remote patient monitoring is defined as a technology that enables patient monitoring outside of conventional healthcare clinical settings. It is the process of using telecommunications or electronic information processing technologies to monitor the health status of a patient remotely. Mobile patient monitoring on the other hand is a continuous or periodic health monitoring of a patient from a distance by employing mobile device and wireless communications technologies. Figure 1 shows the relationship among all the e-health paradigms scope.

Mobile e-health has become the new developmental trends throughout these recent years as a new pattern of healthcare which integrate communication technology with healthcare system to establish strong infrastructure and improve the healthcare services efficiency. It provides systematic data storage with continuous e-health monitoring system. Both the United States and the United Kingdom have started the e-health implementation programmes and campaigns with the aim to initiate the e-health service application for their whole citizens [39]. Similar to most of new application models, the suggestion and utilization of mobile e-health has not yet ascertained and a lot of research need to be done in order to enhance its configuration most importantly in terms of security. The wide range of potential implementations for mobile e-health applications has greatly contributed to the wide acceptance in e-health. Improvement in electronics, sensor networking and communication technology has assisted the current e-health development system and has provided the means to deploy mobile e-health applications in a variety of circumstances. Following

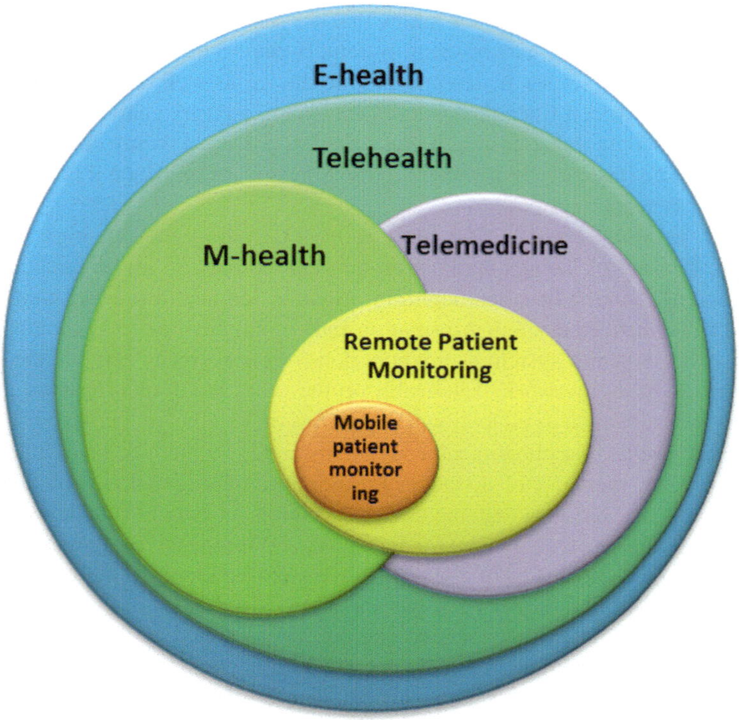

Fig. 1 E-health overall scope

subsections present the details on mobile e-health applications followed by a thorough analysis on the current mobile e-health network architecture and existing mobile e-health authentication protocols. Figure 2 depicts the general overview of mobile e-health and its basic communication system.

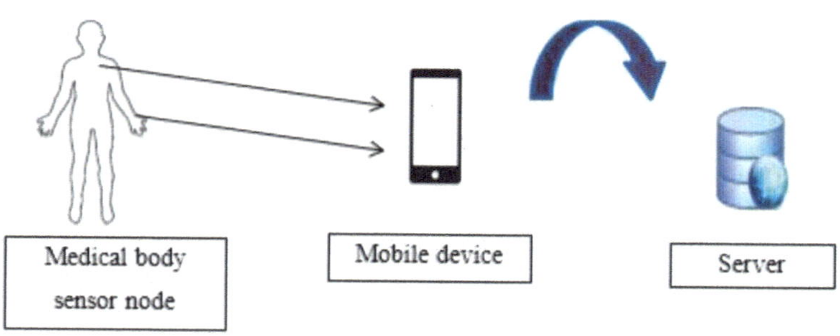

Fig. 2 General overview of mobile e-health system

Mobile E-Health Applications

By integrating mobile devices like smartphones in e-health monitoring system, it does provide a great assistance in patient-doctor interactions since the users are able to perform their own health monitoring without having to visit the healthcare professionals regularly [72]. The mobile device applications will be able to detect the abnormalities from the e-health sensors which deployed in patient body area network (BAN). Rapid development in wireless sensor network together with the recent improvement in sensor networking components such as body sensors have empowered the development of the wireless body area network particularly in healthcare system. However, security loopholes are yet to be eliminated in the e-health security layer in order to protect the integrity and security of WBAN system. Mobile health monitoring system offers great assistance to the healthcare organizations in terms of patient to doctor interaction and it brings potential benefits to the e-health development system. Due to mobile e-health applications and portable e-health sensor nodes development, mobile e-health can be utilized almost everywhere. Also, remote mobile e-health monitoring is now possible to be implemented because of the enhancement in mobile e-health.

Advancements in mobile technology have allowed mobile devices, such as smartphones and tablets, to be used in a variety of different applications. With sensor networking connection and the Internet of Things (IOT) development, as well as widely usage of mobile phones due to its ubiquitous feature, it is now becoming more feasible to utilize mobile technology for medical applications. A user can simply connect a wireless sensor on their body and transmit the e-health data to a mobile device for mobile e-health monitoring [73]. There are a lot of wearable health monitoring applications that has been developed using wireless sensor network. These applications are intended to provide better personal health management and ubiquitous monitoring health system. Furthermore, many countries have shown quite an interest in adapting the mobile e-health system, which is the provision of healthcare services through mobile communications. Due to significant growth in mobile communications in these countries over the past decade, mobile e-health system get the benefits from the mobile communications improvement and avoid the expensive cost for fixed line telecommunication infrastructure in order to adapt this mobile e-health technology. Interest for seamless and effective healthcare service system has increased due to the development of the e-health innovation. People these days have been uncovered with the quick development of portable innovation framework and different applications using mobile devices. Currently, the function of mobile device is not only to communicate but it can also be integrated with sensor networking hence making mobile e-health communication these days much simpler and more reliable. Benefitting from the rapid development in mobile technologies, mobile e-health or m-health initiative has led to the personal health monitoring system which has improved the healthcare current services by the healthcare professionals [72, 74].

There are several existing mobile e-health applications that have been developed recently. CustoMed application [75] is developed for post-knee surgery monitoring using flex sensors on the knees and pressure sensors with the integration of med nodes.

Thilakanathan [73] develop HeartbeatSense mobile e-health application for electrocardiogram (ECG) monitoring and store the health data in cloud computing. In [76], real-time heart monitoring mobile e-health application is developed by implementing a pulse sensor node and then transmit the data for further diagnosis. SecourHealth [77] provides data collection applications in mobile e-health and it is used for remote data collections such as identifying diseases in certain area or tracking people with chronic illness. PHM and MobiHealth applications [16] are developed to monitor cardiac arrhythmia but not implementing any security solution in their applications. Lots of mobile e-health applications [21, 22, 31, 58] focus on developing a pervasive electronic medical record for efficient transmission and data storage where patient identification number need to be inserted in order to access the personal e-health record [21]. CSCW based technique [64, 74] is done to model the cooperation levels among user roles in mobile e-health monitoring activities through system layer. Diabetic telephone monitoring [34] is developed to monitor the blood glucose levels and to update the personal e-health records. In [78], implementation of mobile e-health application to assist elderly who lives alone is developed by incorporating several sensor nodes that can sense their movements around the house. High speed uplink mobile e-health system is developed in [79] to enhance the performance in data transmission and signal processing. In general, lots of research on mobile e-health application currently are being extensively developed which show great importance of mobile e-health nowadays.

However, most of the mobile e-health proposals do not apply adequate security solutions therefore hindering the applications to be applied for real deployment. In mobile e-health application, small and lightweight devices with sensing, computational and communication features need to be unobtrusively deployed on the patient body [78]. Therefore, mobile e-health is strongly constrained by several factors and the design for security procedures should consider the existing constraints in mobile e-health development. These limited constraints are due to sensor node nature itself and wireless channel communication used in mobile e-health as well as the mobile device limited capability.

2.1 E-Health Sensor Node Constraints

E-health sensor nodes in general consists of embedded hardware with low power consumption and small microprocessor with little amount of flash memory. Most security solutions require high computation and large memory usage which then will create a huge challenge when dealing with small e-health sensor nodes. Each of these limitations is due to the physical size of the sensor node itself. Table 1 lists the detail specifications of several different types of e-health sensor nodes that are used in mobile e-health.

Guo et al. [17] uses 2.8 GHz Intel processor and 1GB RAM with Ubuntu platform to develop its mobile e-health authentication protocol. In [44], a prototype for e-health

Table 1 E-health sensor node platform and its specifications

Sensor node	Processor	Data memory/ flash memory	Power (V)	E-health application
Mica2, MicaZ	Atmel 128-8 bit	512 K data flash/128 K FLASH	2.7–3.3	CodeBlue (pulse oximeter sensor, EKG sensor) [80]
Imote2	Marvell PXA271-32 bit	256 kB SRAM/32 MB FLASH	3.2–4.5	Real-time structural health monitoring [81]
TelosB/ TmoteSky	Texas instruments MSP430-16 bit	10 K RAM/ 48 K FLASH	3	Real-time BAN monitoring (motion sensor, BP sensor, ECG sensor) [71, 82]
Shimmer	Texas instrument MSP430F1611-16 bit	10 KB SRAM/48 K FLASH	1.8–3.6	WBSN (waist sensor, accelerometer, heart rate sensor) [83]
AquisGrain	Atmel ATMega128L-8 bit	4 KB RAM/ 512 KB FLASH	1.8–5.5	Pervasive healthcare applications [84]
Arduino	Atmel ATmega328-32 bit	2 KB RAM/ 32 KB FLASH	2.5–12	Real-time heart monitoring and e-health android application (pulse sensor) [76]

application is developed using sensor network platform TmoteSky with 8 MHz TI-MSP430 microcontroller. It has 10 KB RAM and 48 KB Flash ROM. Yang et al. [85] uses ARMv7 microprocessor with 1 GB RAM for their basic identity-based authentication scheme in mobile e-health.

Based on the above findings, computation part in e-health sensor node basically is not excessive as those in other embedded system thus security algorithms should be developed meticulously for e-health sensor network application. E-health sensor node memory basically has a very limited storage capability and therefore makes it impractical to support complex security algorithms for e-health authentication. These e-health sensor nodes have a very limited memory size and limited radio transmission range with data processing capability as well as limited battery power.

2.2 *Mobile Device Constraints*

Mobile e-health application has to consider the constraints in mobile devices as well due to limited resources such as low CPU, low storage capacity and low battery power [24]. Mobile devices come with limited CPU processing and mobile e-health application basically do not support high or complex programming compared to a computer or an e-health server. Therefore, to implement mobile e-health application, security algorithms should be developed uncomplicated and lightweight due to the constraints in the mobile device itself.

2.3 Limited Energy

Energy consumption is a very essential factor that should be put into consideration while designing the authentication security protocols for mobile e-health application. Given the high energy usage of WSN topology and mobile e-health, it is important to limit the power consumption in order to extend the battery lifetime. Additionally, by incorporating security solutions to mobile e-health network, significant energy consumption will be used for instance in performing encryption and decryption process.

2.4 Limited Data Storage

For mobile e-health application which consist of sensor nodes and mobile device operation, data storage capability of these two components are very limited. High and complex application will then reduce the speed and the efficiency of m-health application. Therefore, e-health data and complex security solutions should be handled by the e-health server in order to improve the efficiency of m-health application and at the same time protecting the security of sensor nodes and mobile device as well. Plus, it is crucial to develop lightweight security solutions in sensor node or mobile device to secure the network communication due to its limited memory and storage system capability.

2.5 Network Constraints

Due to the open access channel in e-health network, the communication is inherently unreliable thus making the e-health system to be exposed with security threats and attacks. Besides, sensor networks normally utilize several nodes in the nearby network environment which potentially exposed to packets and data collision. Interference with other sensor nodes network will create a security challenge in securing the network channel. In addition, most current e-health systems transmit their data over a third-party server such as e-health service provider or certificate authority for authentication process which open the security loopholes in the e-health network.

3 Mobile E-Health Network Architecture

World Health Organization (WHO) and m-Health Alliance has highlighted that data security and user privacy is the uppermost issues that should be put more attention in mobile e-health applications [77]. Chan [72] presents a basic network architecture for mobile e-health monitoring system. The network architecture consists of three levels which are the body area network (BAN), the personal area network (PAN), and the wide area network (WAN). However, it proposes multiple users access to the healthcare database without securing the authentication process. Moreover, by turning on the Bluetooth and GPRS connectivity in the mobile device, it has greatly reduced the battery lifetime which is the major problem in this proposed mobile e-health application. Wang et al. [33] then designed an elementary mobile e-health architecture which includes e-health sensor nodes, a mobile phone, a central server and a client-side platform. This network architecture uses only a simple username and password for its authentication.

Mobile e-health current architecture generally is a three-tier architecture that includes a third party server in the network architecture. In [64, 74], it shows that a user needs to transmit the data through his or her mobile phone to a third party server before transferring the data into the medical database. The third party server is basically either the e-health service provider or certificate authority and by expanding the e-health network to a third party server, it opens bigger security vulnerabilities to the e-health network system. In [73], HeartbeatSense network architecture is developed and a trusted third party server is needed for data sharing service. A lot of security issues then are being discussed when cloud storage is being implemented in mobile e-health system [86]. It puts the privacy and confidentiality of the e-health data in high risk if the security authentication scheme for mobile e-health is not being enhanced. Figure 3 shows the architecture of HeartbeatSense that needs a trusted third party for its data storage service. MobiHealth [87] uses a third party Wireless Service Broker (WSB) as a surrogate host to transfer the data and store the data in BANData repository before sending it to end-user in mobile e-health application while Nagaty et al. [24] needs a trusted third party server for security configuration which is to distribute secret key to mobile devices and generates a message authentication code (MAC).

Wireless Sensor Network (WSN) normally will be remotely managed and therefore its potentially threatening environment will expose the system to the security attacks. With the nature of open access channel in WSN, it allows the adversaries to have a greater access to attack the mobile e-health network. It is necessary to control the information access otherwise the data will be exposed to malicious third party and many security threats that can subject it to wrong use. Based on several thorough discussion on security analysis on the mobile e-health, the user secret information especially the verification data should not be exposed to the registration centre or any third party server [88]. A secure and reliable security measures need to be developed in the mobile e-health network to enhance the level of the security system. On top of that, authentication is the most fundamental requirement needed in order to prevent illegal access into the mobile e-health network system.

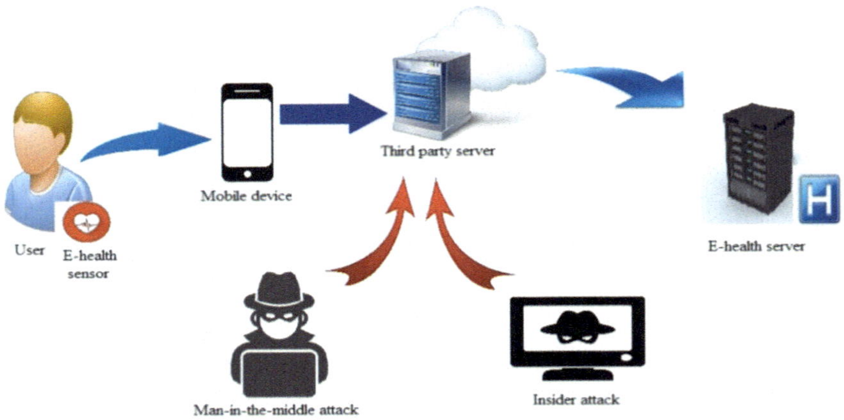

Fig. 3 Security attacks in the three-tier mobile e-health network architecture

Ubiquitous health monitoring (UHM) is then developed through smart WBAN sensor technology mainly to support senior citizens lifestyles. UHM general architecture is developed which consists of WBAN, access networks, and lastly data processing and storage. The e-health sensor collects and transmits the patient data to a personal mobile device where the mobile device provides a display and information processing for secure transmission to a remote e-health server via mobile network or WLAN. Authorized healthcare professional then can analyse the patient data through the e-health server. Due to the open access network in WSN, it is very important to develop an e-health network architecture that limits the communication boundaries within the e-health users only. A third party server or any certificate authority should not be included within the network architecture. UHM architecture proposes direct data transmission to a medical server without going through into any third party server. In order to eliminate the role of the third party server or the certificate authority, identity-based authentication should be implemented since it does not depend on a certificate authority. However the authentication scheme should be improved where the user anonymity should be protected to avoid user impersonation or sensor node cloning attack. UHM architecture has proposed the identity based authentication however the identity of the sensor node and the identity of the mobile device is straightforward and it does not go through any secure process of generating a unique identity.

A typical structure of three-tier e-health system architecture is shown in [44] where the first tier includes the wireless body area network communication in which the e-health sensor nodes are placed in the patient body. The second tier involves e-health data processing between the mobile device and the third party base station. The mobile device will collect and transmit the e-health data through a third party server. Then, the last tier includes data storage in the e-health central server and provides user access to the e-health network. Most mobile e-health network architectures still use a three-tier network architecture that depends on a third party server or e-health service provider either for data storage service or authentication service. In order to protect user anonymity and user untraceability, any third party except the communicating agents cannot knows the identity of the user or get involved in the user data transmission [89]. Therefore, the role for a third party server in mobile e-health network architecture should be eliminated and a new two-tier e-health architecture is developed in this research to facilitate the security requirement in the mobile e-health network. The next Sect. 3.1 will discuss in detail on the rising issues and problems of a third party server implementation in mobile e-health network.

3.1 Three-Tier Mobile E-Health Network Architecture

Network architecture is one of the most important security designs in mobile e-health system. The e-health system enhances and supports healthcare delivery services benefitting from the increased accessibility of communication technologies as well as the advancement of monitoring devices. It is now possible to bring telemedicine directly anywhere anyhow in our daily environment. With the advancement of the current IOT, mobile e-health network architecture has been developed to be more feasible and reliable in data communication system. Mobile e-health network architecture refers to the layout and design of a complete framework of the mobile e-health network consisting of the hardware, software, connectivity and communication protocols. Current three tier network architecture is no longer reliable since it implements a third party server or an e-health service provider to analyse the mobile e-health data before sending it to the e-health server. This will lead to perilous security attacks and put the whole mobile e-health system in risky situation where an insider attack and man-in-the-middle attack could affect the reliability of the mobile e-health network.

This three tier network topology leads to security attack exposure and therefore reduces its reliability throughout the network communication. Network architecture basically is developed to improve the efficiency of the mobile e-health system by implementing essential element in the whole system. Insider attack occurs when an attack is perpetrated on the mobile e-health network by a person with legitimate system access. Insiders will have authorized access to the system and may have specific objectives in the network architecture. For instance, insiders can be the third party server or the e-health service provider as well as the cloud service provider. According to NIST Computer Security Resouce Center [90], this type of attack is extremely difficult to detect and it is hard to protect the system against insider attack. The security issues in a third party server or the cloud and e-health service provider such as insider attack and spoofing attack have been thoroughly deliberated in [12, 91, 92] and Zissis et al. [12] has stated to focus on the trusted authentication scheme implementation as the security solution in the system. Khorshed et al. [51] has also discussed the issues regarding the third party implementation in cloud service provider that leads to unknown risk profile and lack of transparency. Unknown risk profile is happened due to the service provider unwillingness in providing the security logs and audit thus without enough security procedure implementation, the e-health system will be exposed to various unknown risks and it is not easy to deal this issue while maintaining the third party server association in the e-health system.

Besides, a lot of current researches [44, 93–95] also stated that e-health current network architecture requires a very high cost implementation which is one of the major challenges in e-health system. Thus, by improving the current three-tier mobile e-health network architecture to two-tier network architecture can reduce the mobile e-health cost implementation.

3.2 Third Party Server

According to Xu et al. [96], preserving user privacy and transmitting e-health data securely in mobile e-health application should be paid more attention. The authentication schemes should be designed to secure not only the mobile e-health system but to secure the communicating environment as well. Most of the network architecture implements a third party server in e-health system with the main purpose to generate the mobile e-health security parameters since most e-health server in the hospital cannot facilitate immense and complicated security programming due to limited computing resources and limited memory space [36]. Therefore, the need for Certificate Authority (CA) or PKG is to facilitate with the security keys and security parameters generation since most of authentication protocols needs complex coding and thus with e-health sensor node and simply mobile device is not enough to do those keys generation as well as security keys management for mobile e-health system. However, Moghaddam et al. [37] has identified the encryption algorithm issues with the e-health service provider that happen when the user outsources the e-health sensitive data shared on a third party server. With all the security keys stored in the third party server, the mobile e-health system will be affected of losing those private keys in a security attack. According to [13, 39, 62], the e-health service provider role is to achieve maximum reliability which results in full performance in encryption, decryption, and key generation process. Therefore, lightweight authentication protocol should be implemented in the mobile e-health system in order to eliminate these security issues regarding the third party server in the mobile e-health system. With the lightweight authentication protocol implementation in the system, the third party server or the e-health service provider is no longer needed in the current mobile e-health network architecture.

4 Access Control Structure

Access control is a security procedure that can be used to legalize a user whether he or she can view or use the data resource in a networking system. It limits the connections to the system data and server. Access control structure in mobile e-health system is infinitely significant due to open access nature in mobile e-health data transmission. Through access control structure, user identification can be done with proper authentication. Basically, access control system provides critical processes which are authentication, authorization, and accountability. These essential security procedures are needed to implement high security in the e-health access control structure. The main objective of the credential and trust management systems is basically to enable secure access control among users. However, due to distributed nature of WSN particularly in mobile e-health, access control structure is susceptible to various security attacks such as replay attack and node cloning attack.

Mobile e-health is now increasingly considered as a valuable approach to enhance the quality of healthcare system management due to the development of IOT [97]. IOT technology provides a competent and structured approach to remodel the mobile e-health division due to ubiquitous computing nature in the system. Current development in IOT has allowed the mobile e-health to be developed in a more efficient and intelligent way to improve the mobile e-health system capabilities [98]. However, many of the mobile e-health authentication schemes are vulnerable to various types of attacks, inefficient and do not provide user friendly approach especially for the elders [99, 100]. Low computation algorithm should be implemented to facilitate the mobile devices low computation power capability. Since mobile e-health communication is transmitted over the open access network, its access control structure should be well developed to secure the channel transmission. Access control encompasses security techniques that provide selective restriction to the mobile e-health data. It is a way of limiting access to the mobile e-health resources according to the requirements of the whole system. Due to insecure transmission of mobile e-health data over the open access channel such as Internet, security and privacy is the major obstacle that should be addressed when developing the mobile e-health application [89]. The purpose of access control techniques is that only authorized user should communicate with the e-health server and decrypt the transmitted data. Therefore, a secure authentication is needed for the access control structure and its function is extremely important to ensure that the mobile e-health data resources are not available to unauthorized users. Access authorization for data resources in mobile e-health application can be designated with the support from the efficient user access control mechanisms [101]. Only the authorized users with unique access privilege can get through the mobile e-health data system.

Implementing SSL protocol authentication in the mobile e-health network to create a secure connection between a user and e-health server is no longer reliable. SSL is a technology that establishes an encryption between a server and a user, typically using a web server or a website and a browser. The communication between both parties will be slower compared to the communication without SSL since it needs extra work in exchanging handshakes as well as in encryption and decryption process. It can increase the network traffic and reduce the response runtime in the e-health server [100, 102]. Likewise, password-based authentication is also no longer reliable and with a lot of improved schemes that unexpectedly turn out to be vulnerable to security attacks such as parallel session attack [103]. Thorough security issues on password-based authentication protocol will be discussed in the next subsequent section.

5 Existing Mobile E-Health Authentication Protocol

In-depth study has been done throughout this thesis to evaluate and to analyse the mobile e-health authentication as well as the current mobile e-health network architecture. Thilakanathan et al. [73] has proposed email and password authentication and has implemented ElGamal encryption in the authentication protocol. The authentication protocol needs the Cloud Storage Provider to verify the user credentials. However, cloud computing is more susceptible to security and privacy attacks since the data transmitted is highly sensitive and private thus many health organizations are reluctant to implement cloud computing in mobile e-health application. The encryption protocols used is very complex to be implemented in a small and low power mobile e-health application thus it will reduce its feasibility since it takes significantly longer for the mobile device to transmit the data through the proposed HeartbeatSense application. SecourHealth authentication scheme [77] uses username and password for user authentication and it can be done in offline mode. However only for the first authentication, user registration requires online connectivity to the e-health server and the authentication protocol will generate the information required for offline authentications. Lightweight security framework for SecourHealth is developed to secure the data exchange between the mobile device and the e-health server. Besides, the security mechanism provides data protection by securely encrypting the locally stored data and proposes forward secrecy in the authentication protocol. The weakness of this mobile health authentication scheme is the device sharing concept where it allows user access from any mobile devices in which the SecourHealth application have been installed. Therefore, this authentication scheme is exposed to more security attacks where the adversary can hack into the mobile health application using different mobile device since only username and password are needed in order to get authenticated. Its security protocol analysis is not formally validated and thoroughly analysed yet it focuses only on the authentication protocol implementation.

Privacy-Preserving Attribute-Based authentication system [17] authenticates user attributes through decentralized mobile e-health system. Users can authenticate with the healthcare professionals using verifiable attributes while keeping their attributes undisclosed by using attribute-based encryption (ABE). The verification process is done by using the user's attributes such as blood type, allergies, previous medication, etc. while the real identity is hidden to prevent tracing attacks and collusion attacks. However the authentication system requires the involvement from the third party server or a trust authority (TA) and semi-trusted registration centre (RC) in order to generate and issue credential user attributes. Therefore, it opens the security loophole in the authentication model and the security protocol will be exposed to insider attacks. Plus, the computational cost for the mobile e-health system is much higher and it takes longer processing time for mobile e-health data transmission.

Wang et al. [33] proposes an open communication protocol between the e-health sensors and mobile device through Bluetooth transmission. After the e-health sensor node is turned on, it will automatically search the pairing mobile device and get connected through device identity and then a password is needed for the authentication process. E-health data is stored in a central server and then transmitted to the client-side where the health professionals can access and view the data. This authentication protocol also focuses on the implementation basis and not focusing on the security analysis. Due to its open communication protocol with no security protocol enforced to the mobile e-health system, this authentication scheme is vulnerable to replay attack, node cloning attack as well as password-break attack.

Ubiquitous Health Monitoring (UHM) system [35] suggests a mutual authentication between the sensor node and mobile device and then between mobile device and e-health server to prevent impersonation and replay attack. The authentication protocol uses symmetric encryption using the identity of the e-health sensor node and the mobile device identity. However the e-health sensor identity is taken from the device number and it is not generated as a unique identity that cannot be regenerated. Symmetric key algorithm uses identical security keys for encryption and decryption which is no longer secure compared to asymmetric encryption which uses different keys for encryption and decryption. Since e-health sensor node, mobile device, and e-health server share the same secure session key, it will become a security issue if any of the entities is stolen. Li et al. [44] then proposed a group device pairing (GDP) for UHM to provide consistent health monitoring and real time data collection. This authentication scheme is also based on symmetric cryptography. Li, et al. eliminates the function of PKI and proposes an efficient key management system during pre-deployment and deployment phase. However, GDP protocol requires additional computation overhead which will increase the protocol runtime plus the user identity is not securely generated. The same group key is used and distributed over the network will expose the mobile e-health system to the insider attack.

Jones et al. [16, 71] and Marti [87] propose MobiHealth architecture as well as the framework for mobile e-health remote monitoring. However, the mobile e-health framework does not implement any security protocol or authentication within the e-health sensor node and mobile device. Basic SSL authentication in the e-health server is used to establish the connection between the user and the e-health server. The security framework is not securely developed since it is vulnerable to various attacks such as replay attacks, sensor node cloning attacks and man-in-the-middle attacks.

Nagaty [24] proposes identity based authentication using the patient identity and validates the role of the user through Role Based Access Control (RBAC) to control the user limit access. RBAC authentication manages to restrict unauthorized user access in the mobile e-health system but it only uses login and password scheme for security level authentication. The mobile e-health system uses a trusted third party server to generate security parameters and encrypt the e-health data in the cloud storage. However, the proposed authentication protocol does not consider the situation if the mobile device is stolen since the secret token or password will be sent to the mobile e-health application installed on the user mobile device. Plus, the system does not protect user anonymity since patient real identity is used for authentication.

Santos et al. [50] proposed Radio Frequency Identification (RFID) tags for patient identification in mobile e-health application for medication control. By using the automatic user identification, the systematic errors in mobile e-health procedure might be reduced. However, RFID authentication is quite costly to be implemented and the data can only be transmitted if there is enough radio frequency signal strength in the network. Therefore, it is not reliable to use RFID for mobile e-health since the network system should be able to get accessed in any condition. The authentication protocol uses public key certificates that has to be verified by a third party server or a Certificate Authority (CA). Instead of username, the user will insert the RFID user identity and the certificate server will issue a user password during the registration phase and the password will be stored in its database. The proposed authentication scheme is still based on login and password method which is not secure anymore and it does not protect user anonymity. In addition, RFID authentication is quite a complicated and complex procedure to be implemented in the proposed mobile e-health monitoring that targeted the elderly people as the common users. Analysis on RFID authentication schemes in mobile e-health has been discussed in [48] and most of the schemes do not satisfy the security requirements in e-health and have very low performance. Many security attributes cannot be fully implemented in basic RFID authentication scheme. He et al. [48] has shown that most RFID authentication schemes are vulnerable to tag impersonation and forward secrecy attack since the adversary might retrieve the secret information in the tag such as the user identity and password.

Wang et al. [33] and He et al. [104] proposes a low cost and flexible mobile e-health application. Wang, et al. uses Bluetooth to transmit the data from the e-health sensor nodes to the mobile device. However, they only apply username and password in their authentication schemes without any additional security measures which will expose the system to eavesdropping and password-break attack. For that reason, Wu et al. [105, 106] then introduces a two-factor authentication in mobile e-health where a username and password with additional smart card implementation is used to enhance the security of mobile e-health system. Smartcard based password authentication requires a user to have a valid smartcard with a correct user password. Wu, et al. proposes the authentication scheme focusing on protecting the electronic patient records in the e-health data system. However, He et al. [88] has shown that Wu, et al. authentication scheme is exposed to user impersonation attack and insider attack. Since Wu, et al. implement a third party server in their security protocol and use the same password to access several servers, the privileged insider of the third party server might learn the user password and impersonate the user. Wei et al. [107] also shows that Wu, et al. authentication scheme as well as He, et al. authentication scheme are not secure and they are susceptible to replay attack besides the password break attack and smart card attack. Wei, et al. then improved the authentication scheme by eliminating the pre-computing phase in Wu, et al. and He, et al. authentication protocols that is said to be the main cause for the security attacks. Pre-computing phase is used to settle the complex computation process in the authentication system and reduce the time consuming to extract the security values in the security protocol. Hence, Wei, et al. system takes longer time for security computation during the authentication phase since mobile device does not have large memory space and it has low computational abilities. On the other hand, Zhu [108] has showed that Wei, et al. system however is vulnerable to offline password-break attack and the adversary has total control over the communication channel between the user and the server in the login and authentication phases. Besides, Wei, et al. authentication scheme does not be able to protect user anonymity and forward secrecy attack. Zhu then proposed an improved authentication scheme to overcome those weaknesses in Wei, et al. scheme. Later, Khan et al. [109, 110] and Lee et al. [111] have showed that Zhu authentication scheme is not secure and it is still vulnerable to password break attack besides user traceability and denial-of-service attacks. Khan, et al. then proposes a secure two-factor authentication protocol based on Diffie-Hellman key exchange method. However, the authentication scheme transmits plaintext user identity over the communication channel which does not provide user anonymity. Plus, the system has higher computation cost on e-health server compared to Wei, et al. authentication scheme and Zhu authentication scheme.

In [85], Yang, et al. stated that the exposure of secret keys through mobile device brought up security concerns in its cryptosystem therefore key evolving technique is proposed where the private keys will evolve per period. However, key evolution technique has not been well developed due to the difficulty to set a proper and exact time period range for the key evolution and the system needs another third party server to assist the key update. Key evolution technique for parameters such as private key is quite complicated to be implemented. Unique identity is indeed an efficient way to strengthen the security in mobile e-health but Yang, et al. authentication scheme uses only pseudonym of the real patient identity in their authentication protocol. Plus, no security protocol is being implemented in the e-health sensor node within the body area network which will expose the mobile e-health system to sensor node cloning attack.

Abundant of research has been done in order to develop a secure authentication scheme in mobile e-health however most research implement password-based authentication in their security protocol [60, 88, 89, 97, 104–109, 111–115]. There are several of the authentication schemes that use smartcard implementation together with username and password authentication in order to improve their security level. However, password-based authentication is already proven to be no longer secure since it is vulnerable to password break attack and insider attacks plus the smart card might be duplicated, lost or stolen as well [116]. In a two-factor authentication scheme, there are basically four main stages in the security protocol which are registration phase, login phase, authentication phase and then password change phase. Nevertheless, password-based authentication protocols generally do not provide strong user anonymity since username or real patient identity is being transmitted over the public channel and the system is highly vulnerable to user impersonation attack as well as forward secrecy attack. Many security loopholes have been identified in this kind of authentication scheme which makes mobile e-health application is not reliable to be implemented due to high privacy and sensitive data in healthcare.

A common feature among username and password based authentication is that the user identity is static in every communication session which may reveal the user information and it is exposed to user impersonation or identity theft attack. Wang et al. [112] then proposed a dynamic ID based authentication in their username and password authentication scheme to protect the user anonymity. However, Khan et al. [113] has proved that Wang, et al. scheme is not practicable since it does not provide user anonymity and it is vulnerable to insider attack. Then, Khan, et al. proposed an improved dynamic ID based authentication by adding a secure session key in their security protocol. Chen et al. [97] then has proved that Khan, et al. scheme is vulnerable to insider attack and its password change phase is not verified. Chen, et al. then minimized the security algorithms from Khan, et al. authentication protocols to improve the efficiency of the mobile e-health system and provide user anonymity. Chen, et al. authentication scheme however is vulnerable to replay attack since the adversary can eavesdrop the communication from the user to the e-health server within the timestamp. Besides, it is vulnerable to insider attack since the secret key is shared by all the legal users. Xie et al. [114] and Jiang et al. [89] then proved that Chen, et al. authentication scheme does not provide user anonymity and it is vulnerable to impersonation attack. Jiang, et al. then proposed an enhanced authentication scheme in order to protect user anonymity. Wu et al. [115] however has proved that Jiang et al. authentication scheme is on the other hand vulnerable to user impersonation attack.

Tan [117] and Awasthi-Srivastava scheme [118] then proposes a three-factor authentication scheme for mobile e-health which combines password-based authentication scheme with the use of smart card and biometric authentication. However, since the identity of the user is transmitted in the form of plaintext, Tan [117] has proved that Awasthi-Srivastava scheme is susceptible to reflection attack and the system does not protect user anonymity. Tan then proposes a modified authentication scheme from Awasti-Srivastava by adding a secure session key in the authentication phase. Also, the user identity is not transmitted in any plaintext form to provide user anonymity. Tan authentication protocol applies elliptic curve cryptosystem (ECC) which needs smaller key size compared to other authentication resulting Tan authentication scheme to be low power consumption and higher speed utilization. However, the password sent to the e-health server is not encrypted which will make the mobile e-health system vulnerable to password-break attack and Arshad et al. [116] then shows that Tan authentication is insecure against replay attacks and denial-of-service attacks. Both Tan and Awasti-Srivastava schemes are costly and use high computation memory since the authentication schemes are using smart card and complex algorithms in their authentication protocols. Arshad et al. [116] then proposes a modified ECC-based three factor authentication for mobile e-health to improve the security and efficiency of the Tan authentication scheme. Arshad-Nikooghadam scheme uses a timestamp and fresh random numbers to eliminate replay attack. The authentication protocol consists of four phases which are setup phase where the e-health server will generate security parameters and a master secret key to be used in the registration, login and authentication phases. Both Tan and Arshad-Nikooghadam schemes has reduced the computational cost significantly with the implementation of ECC.

However, Das et al. [119, 120] has showed that Arshad-Nikooghadam authentication scheme fails to protect insider attack and the system does not provide formal security analysis. Das, et al. also shows that Tan authentication scheme fails to provide proper authentication and the system is vulnerable to replay attack. Das, et al. then proposes a solution to Tan and Arshad-Nikooghadam authentication schemes and provides formal security analysis using AVISPA tool. In the proposed authentication protocol, Das, et al. does not directly store the user password and biometric information into the smart card in order to avoid insider attack. However, the computational cost for Das, et al. scheme implementation is much higher compared to Tan and Arshad-Nikooghadam schemes and a third party server is needed to generate the security parameters.

Chatterjee and Das [101] then propose a new password-based authentication protocol with attribute-based encryption based on elliptic curve cryptography (ECC). The mobile device should have a private key with matching set of user attributes in order to retrieve the encrypted data transmitted over the network. However, their authentication scheme still use password based authentication and smart card implementation focusing only on sensor networking. Therefore, the mobile e-health system is vulnerable to password-break and replay attack. Plus, the authentication protocol is not been applied in term of real mobile e-health application.

Table 2 shows in details on the existing mobile e-health authentication protocol and the thorough analysis on each protocol.

6 Security Issues in Mobile E-Health Authentication

Since mobile e-health application is an open communication channel, security and privacy issues are the most worrying obstruction in its development. Therefore, authentication schemes are very important in mobile e-health application since it is the first line of defence in every network application [97, 107, 108, 112, 117]. It is important to establish secure and trust relationship among the e-health devices to provide secure and efficient initialization in e-health communication system [44].

6.1 Username and Password Login Phase

Mobile e-health application should apply highly secure and strong security procedures to avoid false treatments, incorrect identifications, and unauthorized access. Most early authentication schemes in mobile e-health are merely based on username

Table 2 Existing mobile e-health authentication protocol

Previous work	Description	Drawbacks	Authentication scheme	Involvement of third party server	Formal security analysis
Thilakanantan et al. [73]	Implement ElGamal encryption and develop HeartbeatSense application	Complex authentication protocols Low feasibility Longer time to transmit the e-health data	Username and password	Yes Needs cloud storage provider to verify user credentials	No
Simplicio et al. [77]	Lightweight security framework for e-health data exchange Offline authentication can be done	Exposed to security attacks such as hacking and eavesdropping since user access can be done from any mobile devices	Username and password	Yes Needs third party server to generate security parameters	No
Guo et al. [17]	Authentication is verified using user attributes to keep the real identity hidden Implement attribute-based encryption and ECC	Involvement of two third party servers which will expose the system to insider attacks and man-in-the-middle attack High cost and takes longer processing time	Attribute-based authentication	Yes To generate and issue user attributes as well as for key distribution	No
Wang et al. [33]	Open communication protocol, automatically search the pairing mobile device through Bluetooth	No security analysis implementation Vulnerable to replay attack, node cloning, and password-break attack	Device ID and password	Yes Implement a central server for data storage	No

| Previous work | Description | Drawbacks | Authentication scheme | Involvement of third party server | Formal security analysis |

(continued)

Table 2 (continued)

Previous work	Description	Drawbacks	Authentication scheme	Involvement of third party server	Formal security analysis
Yeh et al. [35]	Ubiquitous health monitoring system Implement symmetric encryption using the sensor node and mobile device identity	Sensor identity is not unique and it is taken from the device number Exposed to device stolen attack	Device ID	No	No
Li et al. [44]	Propose group device pairing (GDP) protocol Does not rely on PKI	The user identity is not securely generated The same group key is distributed in the network and it is vulnerable to insider attack	ID and password for the controller	Yes For data storage	No
Jones et al. [16, 71] Marti [87]	Proposes MobiHealth architecture Remote monitoring and treatment services	No security protocol is being implemented within the sensor node and mobile device	SSL authentication	Yes Back-end server	No
Nagaty [24]	Implement RBAC to control user access	Does not provide user anonymity Secret token and password attack	Login ID and secret token	Yes Generate security parameters and to encrypt data in cloud	No
Santos et al. [50]	Use of RFID technology in m-health Proposes remote medication control system for elderly people	RFID is costly and requires complex procedure Troubleshoot with the RF low signal link Does not protect user anonymity	RFID tag, user ID and password	Yes Requires certificate authority (CA)	No

(continued)

Table 2 (continued)

Previous work	Description	Drawbacks	Authentication scheme	Involvement of third party server	Formal security analysis
Wang et al. [33] He et al. [88, 104]	Proposes low cost and flexible mobile e-health application Transmit e-health data through Bluetooth	No additional security measures Vulnerable to eavesdropping and password-break attack	Username and password	Yes Central server is required for e-health database	No
Wu et al. [105, 106]	Two-factor authentication To secure the e-health patient records	Vulnerable to user impersonation attack, insider attack, replay attack, and password break attack	User identity, password, and smartcard	Yes For password verification	No
Wei et al. [107]	Eliminates the pre-computing phase	Longer runtime security computation since pre-computing phase is eliminated Does not protect user anonymity and forward secrecy attack	Username, smartcard and password	Yes For password verification	No
Previous work	Description	Drawbacks	Authentication scheme	Involvement of third party server	Formal security analysis
Zhu [108]	Uses two-factor authentication	The system is vulnerable to denial-of-service and password break attack	Patient identity and password	NA	No
Khan et al. [113]	Uses a secure Diffie-Hellman key exchange method	Transmit plaintext user identity which does not provide user anonymity Higher computation cost	Username and password	NA	No

(continued)

Table 2 (continued)

Previous work	Description	Drawbacks	Authentication scheme	Involvement of third party server	Formal security analysis
Yang et al. [85]	Proposes key evolution technique in m-health Identity-based authentication	Not fully developed due to the difficulty setting a proper time period range for key evolution Vulnerable to sensor node cloning attack	Pseudonym of patient identity	Yes Involvement of helper key update (HKU)	No
Wang et al. [112]	Proposed a dynamic ID based authentication	Does not provide user anonymity and vulnerable to insider attack	Password based authentication	NA	No
Chen et al. [97]	Minimized the security algorithms from Khan, et al. protocol to improve the efficiency of the mobile e-health system	Vulnerable to replay attack and user impersonation attack	Username and password	No	No
Awasthi and Srivastava [118] Tan [117]	Proposes a three factor authentication scheme	Insecure against replay attack costly and high computation memory usage	Username and password, smartcard and biometric	No	No
Arshad et al. [116]	Proposed a modified ECC based three factor authentication Uses a timestamp and fresh numbers to eliminate replay attack	Fails to protect insider attack	Username and password, smartcard and biometric	No	No
Das et al. [119, 120]	Does not directly store the user password and biometric information in the smart card	Higher computational cost Vulnerable to password break attack and replay attack	Username and password, smartcard and biometric	Yes Trusted authority is needed to generate the security parameters	Yes Provides formal security analysis using AVISPA

and password that will inevitably subject to many security attacks [107]. Smartcard based password authentication schemes are then being introduced to improve the mobile e-health authentication protocol. However, two-factor authentication scheme which includes the smartcard implementation in its protocol is no longer reliable and it is insecure against insider attack and smart card stolen attack. If the smartcard is stolen, an adversary can impersonate a valid registered user if there is no revocation system in the authentication scheme [113]. There are many security drawbacks are found when using the smart card authentication scheme since the adversary can extract all the information stored in the smartcard for instance by mounting power consumption analysis, namely side channel attacks [114].

Implementation of biometric authentication do provide a secure communication however Arshad et al. [116] and Das et al. [119] has showed that the biometric pattern belonging to the same person may vary slightly from time to time hence it will deny the user verification during the authentication phase with just a slight difference. The registered biometrics for instance iris, face, retina, fingerprint etc. may never hold the same value since biometric authentication is highly sensitive. As a result, it is vulnerable to denial-of-service attacks during the login and authentication process due to its extremely sensitive property. Moreover, biometric authentication is not always applicable since it cannot be applied to the sensor devices in the user body area network [44]. Due to the biometric highly sensitive issue in the authentication protocol, this research proposes that sensor node identification should be used instead of user identification.

6.2 Sensor Node Cloning Attack

The sensor networks is left unattended by the healthcare professionals after being deployed in the mobile e-health applications which will then allow the adversary to get greater access to attack and tamper with the sensor networking. Besides, the mobile e-health sensor networking is being remotely managed and thus it is quite difficult to detect any tampering or sensor node cloning attack from the adversaries. Redeploying cryptographic and security parameters are also impractical to be implemented remotely. Therefore, a strong and trustworthy security protocol should be applied in the mobile e-health networking to protect the system from these security attacks. Securing the sensor node is the main security challenge in protecting the mobile e-health network infrastructure as well as the data confidentiality to ensure a secure communication between the patient and the healthcare professional.

Therefore, ensuring that only authenticated and genuine e-health sensor nodes exist in the mobile e-health network is necessary to avoid any interfering from the other entities. It is crucial since the adversaries can introduce fake sensor nodes similar to the authenticated one in the mobile e-health network. Besides, the software running in the e-health sensor nodes might also be tampered hence invalid software will lead to false data and consequently generate false alarm to the mobile e-health system. Moreover, most authentication system can only detect the existence of a fake e-health sensor node in the mobile e-health system after a certain period of time. Subsequently, sensor node cloning attack may already harm the system and cause network disorder by the time the security protocol identifies the attack.

6.3 Reliability of the Mobile E-Health System

Most mobile e-health system depends on a third party server to systematically analyse medical records assessment, unauthorized user access, data modification, system performance, and security parameters generation which will make the mobile e-health system unreliable. Many security and privacy attacks in fact happen within the third party service provider itself since they have direct access to the e-health data storage and some of them even expose and sell the private data in order to gain profit [73]. Moreover, most of the mobile e-health systems nowadays adopt cloud computing for e-health data storage and it has raised many concerns from the health organizations due to the security and privacy issues [91, 93, 121]. Many professional organizations are not interested to use mobile e-health application since the data centres are managed by the other third party providers and not owned or managed by the user itself [24].

According to the rules provided by Health Insurance Portability and Accountability (HIPAA) standards, the user data can be only read and accessed by authorized health personnel [35]. Thus, any mobile e-health system that allows malicious and unauthorized access into the system is in violation of HIPAA requirements.

7 Related Work on Identity Based Authentication

Identity based authentication protocol is developed based on key agreement protocol that was proposed by Diffie and Hellman [122] in 1976. In 1984, Shamir [123] introduced the concept of identity based cryptosystem in which the user public key is a calculated function of his identity such as identification number or social security number. It is based on a public key cryptosystem in which the user chooses his name, email address, etc. as his public key in order to uniquely identify the user. Shamir original objective for identity based encryption was to simplify certificate management. Later in 2001, Boneh and Franklin [124] proposed a fully functional identity-based encryption scheme based on the computational Diffie-Hellman assumption to ensure a secure communication. Moreover, user authentication is more secure and viable through a public key infrastructure implementation. It is important to retain low energy consumption for mobile e-health monitoring applications since heavy and complex security computation will increase the authentication protocol runtime which is not acceptable in mobile e-health especially during emergency period. Several applications on identity-based authentication will be discussed in this subchapter.

Hölbl and Welzer [125] proposed on two-party identity-based key agreement protocol in order to eliminate impersonation attack using Diffie-Hellman algorithm. Identity-based infrastructure works when a public key of a user is derived from their identity. However, Huang et al. [126] has proved that Holbi and Welzer protocol is not secure and it is vulnerable to man-in-the-middle attack as well as impersonation attack. Huang, et al. then proposed an enhanced Holbi-Welzer identity-based protocol by adding a multi-channel security protocol to improve the certificate management in e-health personal area network using public key infrastructure (PKI). The data is transmitted via wireless radio channels and NSB channels so that any attacks that try to block, modify, replay, or fake the transmitted data will be discovered. However, the proposed protocols do not provide user anonymity since the identity transmitted is not encrypted and it might be exposed to insider attacks. In addition, Li et al. [44] has stated that PKI implementation in the mobile e-health sensor networking is no longer practical since it requires high cost infrastructure due to high complexity in e-health sensor nodes revocation process. Li, et al. proposes a group device pairing authentication that uses a shared group key and individual secret keys based on symmetric cryptography. The authentication scheme also implements identity-based authentication and use sensor node identity for verification. The identity is used to prevent a replay attack where Li, et al. includes the sensor node identity together with Diffie-Hellman public key value in the authentication protocol. However, the sensor node identity is not securely and uniquely generated by a secure process. Plus, the authentication protocol generates a group key in order to authenticate the data in a group key agreement scheme which is highly vulnerable to insider attacks.

Yeh et al. [35] proposes identity-based authentication based on device number on the e-health sensor node and mobile device. However, it does not propose unique identity of the sensor node and the authentication scheme is vulnerable to data stolen attack if any of the devices is stolen. Yang et al. [85] proposes identity based authentication with key evolution technique and implement the use of the helper HKU to generate the security keys. The security protocol requires high key security and renewal cost besides difficulty to set a proper time range for the evolution technique. It is yet a complicated and complex procedure to be implemented in mobile e-health that supposed to be user-friendly and easy for the elderly people as a targeted user. In addition, Yang, et al. authentication protocol does not implement any security access control of the e-health sensor node which will expose the system to sensor node cloning attack (Table 3).

8 Security Requirements in Mobile E-Health Authentication

Mutual authentication is the most essential security requirement since it is difficult to determine who is communicating within the network system [35]. Therefore, it is very important to authenticate each level in mobile e-health network. Ramon Marti

Table 3 Related works on identity based authentication

Identity based authentication	Description	Drawbacks	Unique identity	Requires no PKI
Hölbl and Welzer [125]	Public key of a user is derived from its identity	Vulnerable to man-in-the-middle attack and impersonation attack	No	Yes
Huang et al. [126]	Basic identity-based authentication	The identity transmitted is not encrypted and it does not provide user anonymity	No	Yes
Li et al. [44]	Basic identity-based authentication using symmetric cryptography	The e-health sensor node does not securely and uniquely generated	No	No Propose group device pairing authentication
Yeh et al. [35]	Basic identity-based authentication ID = arbitrary string	The identity of the e-health sensor node and mobile device is directly taken from the device number	No	Yes
Yang et al. [85]	Basic identity-based authentication with key evolution technique ID = random string	Key evolving technique requires high security and renewal cost with complex protocol implementation	No	No Required helper HKU to generate security keys

[87] has thoroughly discussed the security requirements that should be implemented in mobile e-health applications. The importance of security requirement standards in mobile e-health implementation is to enable a seamless and feasible communication within a secure networking.

8.1 Access Channel

Access security includes all hardware devices in mobile e-health system. Data from mobile device must be encrypted throughout the communication channel. However, lightweight cryptography for encryption and decryption protocol needs to be developed in order to facilitate the restrictions and constraints in e-health sensor node and mobile device application. The communication protocols need to be lighter and simpler to avoid lengthy communication and the protocol has to be efficient in terms of overhead and bandwidth due to the low voltage power in mobile e-health devices.

8.2 User Authentication

User authentication is the major security requirement that should be implemented in mobile e-health. Strong user authentication is needed to protect the security layer in mobile e-health which includes application layer, transport layer, network layer, and data link layer. It is a process of validating the truth identity and the genuineness of a user to ensure the reliability in the e-health data transmission. A secure authentication scheme between the user and the e-health server is the most principal requirement that is needed to protect data integrity and user confidentiality [89, 97, 114, 115]. The mobile device and the e-health server need to confirm the authenticity of the user or the e-health sensor node through unique identity or unique keys such as username and password. However, since mobile device basically has limited computational ability, it is highly crucial to improve the efficiency of e-health authentication without interrupting the limitations. For authentication protocol, there are basically two essential security goals which are secrecy and authenticity [44]. Secrecy is achieved when a passive adversary such as eavesdropper has no or negligible access to the security keys in the authentication protocols. Authenticity focuses on defining and proving the security of the authentication protocols followed by its security against the adversary.

8.3 Integrity and Data Storage

Security in data link storage provides data protection in order to store the vital data in a secure manner. Implementation of system cloud in data storage might expose the privacy of the e-health data since a third party server provider is involved in the network architecture. Data integrity can be protected through a secure communication message using message authentication code (MAC) that needs a secret shared key between the user and the e-health server. E-health data is the most important asset that needs to be protected through the e-health security protocol and the integrity check of the received data can be ensured using cryptographic services.

8.4 Confidentiality

Confidentiality of the mobile e-health system allows the user to confirm that only the designated receiver has the privilege to access the transmitted data and it should not be exposed to any adversary including the eavesdroppers. Data confidentiality in WSN normally can be achieved through encryption process in the security protocol using the security algorithm implementation. The demand for data confidentiality is highly required and a sensor node for example should never expose its information in order to avoid sensor node cloning attack or node impersonation attack. In the mobile e-health application primarily, the adversary might inject other fake e-health sensor nodes into the network system to interfere with the communication system and therefore sensor node confidentiality will prevent the adversaries from getting any information regarding the e-health sensor node. It is critical to protect the e-health data from unauthorized access, eavesdropping, replay attack and data modification. Moreover, establishing confidentiality in mobile e-health system is truly important most specifically during security keys distribution process in order to create a secure networking environment in the whole mobile e-health system. Therefore implementing security solutions in mobile e-health applications is extremely significant to ensure the system to be reliable and trustworthy.

9 Formal Analysis and Model Checking

Most security protocols are developed to establish either a cryptographic key or achieving secure authentication in the presence of malicious user or an intruder. These recent years, several methods for analysing security protocols have been developed. Formal verification and model checking of system protocol starts by describing the protocol in certain language and then comparing the performance or behaviour of the protocol with a specification of a required behaviour [127]. A verifier then will generate states and if any inconsistency is detected, a diagnosis report on how the problem occurs will be generated. Model checking is one of the verification mechanisms used in the security analysis to verify a new security protocol. It is an automatic verification technique used for software and hardware verification. Protocol design is a subtle invention since there is a high possibility of resulting errors thus making the testing and debugging extremely challenging. Subsequently, formal verification of protocols has encouraged high interest and these methods generally have been applied in most communication and network protocols [127]. One of the advantages of model checking analysis over other security techniques such as simulation and theorem proving is the newly designed protocol will be exhaustively verified in every single execution to ensure absolute and secure performance of the system in a systematic way. Moreover, model checking allows functionality test of the security protocol design at the earliest stage of its development hence reducing the cost and time of product development.

Formerly, several model checking tools has been developed such as SPINS [128], Uppaal [129], and Murϕ [127]. SPINS basically is a general verification tool that support functional system properties such as design and verification of asynchronous process system. SPINS has been used to verify data confidentiality, integrity, and authenticity in Sensor Network Encryption Protocol (SNEP). In SPINS, the designed protocol needs to be developed using PROMELA language and the correctness statement will be in the form of Linear Temporal Logic Syntax. SPINS was modelled to prove the protocol accuracy and its specifications. Murϕ or Murphi verification system is a protocol verification tool in the form of finite-state machine that uses simple high-level language in order to describe the rules, conditions and overall variables [130]. It has several features in detecting design errors such as deadlocks, error, and invariant checking. Murϕ can also be applied to verify link level communication protocol. Next, Uppaal is an application used for modelling, simulation and real time system verification based on on-the-fly searching technique. It is designed for basic invariant and reachability properties by exploring the state space of a system. Uppaal consist of three main parts which are a description language, a simulator and a model-checker. It provides both graphical interface and textual description for automated analysis of system behaviour [129].

However, it is preferable to utilize verification tools that are reserved for security protocols verification since it has more advantages and benefits compared to other verification tools. These include Casper [131], ProVerif [132], and AVISPA [133]. Casper is a security protocol analyser that utilized process algebra CSP and model checker FDR in order to discover a number of attacks on security protocols. The security protocol is defined as an input text file and then, the output will be interpreted through FDR checker. Xin et al. [126] implement Casper verification tool to analyse their newly designed identity based security protocol for wireless personal area network (WPAN) in mobile e-health. The security of this authentication protocol is formally verified using Casper in order to discover the security attacks on the system.

ProVerif is based on applied pi calculus and it supports many cryptographic algorithms such as symmetric and assymetric encryption, and can be used to analyse the security and authentication of cryptographic protocols automatically. Dong et al. [43], Xie et al. [114] and Shehzad et al. [134] has used ProVerif verification tool to analyse their mobile e-health authentication scheme through formal analysis. Cryptographic functions including hash function, digital signatures as well as encryption process can be modelled through this verification tool.

AVISPA [133, 135] on the other hand is a widely accepted formal analysis method and it supports modular and expressive formal language for identifying protocols and their security properties. It is considered as a push-button tool for automated validation of Internet security protocols and it consists of four back-ends which are on-the-fly model checker (OFMC), constraint-logic-based attack searcher (CL-Atse), SAT-based model checker (SATMC), and Tree Automata based on automatic approximations for the analysis of security protocols (TA4SP). The proposed protocols models are developed through High Level Protocol Specification Language (HLPSL) based on roles and the intruder is modelled using Dolev-Yao [136] threat model. Das [119], Chatterjee et al. [101] and Goswami et al. [137] use AVISPA in their security protocols analysis to verify their mobile e-health authentication protocols. Amin and Biswas [138] also implement AVISPA to formally validate their mobile e-health security verification from replay attack and man-in-the-middle attack. In addition, the use of AVISPA as a model checking tool has successfully verified several WSN related protocols such as Sensor Network Encryption Protocol (SNEP) [139] and TinySec [140]. Odelu et al. [141] use AVISPA to validate the security analysis in their authentication protocol from impersonation attack, user anonymity, and man-in-the-middle attack as well as modification attack. The advantages of using AVISPA as a formal model checking are expressive specification language with flexible definition of several behaviours in only one role definition. Since AVISPA has four back-ends tools and each of the back-ends uses different model checking techniques, AVISPA provides more complete result and thorough security analysis on the security protocols. For that reason, this research chooses AVISPA as the model checking tool for the security analysis of the proposed mobile e-health new authentication protocol.

10 Conclusion

This chapter presents in sequence the important matters that should be analysed before developing a framework of mobile e-health authentication protocol. The concern is how to implement the e-health sensor node identity-based authentication and whether it is safe to apply the security protocol on the unsecured platform considering the nature of wireless sensor network which is commonly exposed to physical and software security attack. By first identifying the limitations that exist in the e-health sensor node network followed by an analysis on security issues that arise in mobile e-health authentication which are the vulnerabilities of the mobile e-health system, the threats and the security attacks on the mobile e-health communication, this chapter continues to identify main security requirement and further solution on confirming security in mobile e-health. The fundamental issues that need to be addressed are security and privacy which demand a secure authentication scheme to protect the mobile e-health system [107]. Additionally, this chapter presents a thorough discussion on related works to support the proposed research by providing a detailed overview of works most directly relevant to the selection of security protocols, techniques and methods used in the development of the security framework for mobile e-health authentication protocols.

Through the review and analysis of closely related works, there exist rooms for improvement and security loopholes to be resolved especially in protecting the e-health sensor nodes sensitive credentials from security attacks and provision of remote authentication in mobile e-health sensor network. Although SecourHealth [77], MobiHealth [87], E-SAP [18] etc. are developed to enable secure authentication in mobile e-health, the implementation still inherits few drawbacks from the user anonymity and replay attack issues. In ensuring secure exchange of user sensitive credential information that usually happens at the beginning of e-health sensor node deployment, a secure authentication protocol which is based on sensor node identity-based cryptosystem is proposed. Another security issue in mobile e-health network is the node cloning attack or cloned user identity which leads to reliability concern in the e-health network system. Therefore, to overcome this security issue, this research proposes a non-regenerated unique identity for the e-health sensor node based on the node platform specific values and abolish the use of common authentication scheme which is the username and password. This research also eliminates the involvement of a third party server in the mobile e-health network architecture by implementing a secure cryptographic scheme in the e-health server.

In summary, this chapter provides the justifications and the rationale for each of the techniques and methods used throughout the development of the authentication scheme in mobile e-health. These are adoption of sensor node identity-based encryption instead of using user identity, elimination of a third party server in mobile e-health current architecture, and finally a new passwordless authentication protocol implementation in securing the communication in mobile e-health network.

References

1. Chen, C.-M., Chen, Z., Das, A.K., Chaudhry, S.A.: A security-enhanced and ultralightweight communication protocol for internet of medical things. IEEE Internet Things J. **11**(6), 10168–10182 (2024). https://doi.org/10.1109/JIOT.2023.3327322
2. Desai, N., Shahnasser, H.: A light review of data security and privacy approaches applicable to e-health systems. In: International Conference on Computing Technology and Information Management, pp. 362–367, Dubai (2014)
3. Khan, J.Y., Yuce, M.R.: Wireless body area network (WBAN) for medical applications. J. Med. Syst. **36**(3), 1441–1457 (2012)
4. Ghani, M.K.A., Bali, R.K., Naguib, R.N.G., Marshall, I.M., Shibghatullah, A.S.: The design of flexible front end framework for accessing patient health records through short message service. In: 2007 Asia-Pacific Conference on Applied Electromagnetics, pp. 1–5. IEEE, Melaka (2007). https://doi.org/10.1109/APACE.2007.4603885
5. Boonyarattaphan, A., Bai, Y., Chung, S.: A security framework for e-health service authentication and e-health data transmission. In: International Symposium on Communications and Information Technology (ISCIT), pp. 1213–1218. IEEE, Incheon (2009). https://doi.org/10.1109/ISCIT.2009.5341116
6. Rosales, M.S., Garcia, G.G., Sanchez, G.D.: Efficient message authentication protocol for WSN. WSEAS Trans. Comput. **8**(6), 895–904 (2009)
7. Khan, S.U., Pastrone, C., Lavagno, L., Spirito, M.A.: An energy and memory-efficient key management scheme for mobile heterogeneous sensor networks. In: International Conference on Risks and Security of Internet and Systems (CRiSIS), pp. 1–8. IEEE, Timisoara (2011). https://doi.org/10.1109/CRiSIS.2011.6061832
8. Mair, F.S., May, C., O'Donnell, C., Finch, T., Sullivan, F., Murray, E.: Factors that promote or inhibit the implementation of e-health systems: an explanatory systematic review. Bull. World Health Organ. **90**(5), 357–364 (2012). https://doi.org/10.2471/BLT.11.099424
9. West, D.M., Miller, E.A.: The digital divide in public e-health: barriers to accessibility and privacy in state health department websites. J. Health Care Poor Underserved **17**(3), 652–667 (2006)
10. Saleem, K., Derhab, A., Al-Muhtadi, J.: Low delay and secure M2M communication mechanism for eHealthcare. In: 2014 IEEE 16th International Conference on e-Health Networking, Applications and Services, Healthcom 2014, pp. 105–110 (2015). https://doi.org/10.1109/HealthCom.2014.7001822
11. Van der Linden, H., Kalra, D., Hasman, A., Talmon, J.: Inter-organizational future proof EHR systems. A review of the security and privacy related issues. Elsevier Int. J. Med. Inform. **78**(3), 141–160 (2009). https://doi.org/10.1016/j.ijmedinf.2008.06.013
12. Zissis, D., Lekkas, D.: Addressing cloud computing security issues. Elsevier Future Gener. Comput. Syst. **28**(3), 583–592 (2012). https://doi.org/10.1016/j.future.2010.12.006

13. Fengou, M., Mantas, G., Lymberopoulos, D., Komninos, N., Fengos, S., Lazarou, N.: A new framework architecture for next generation e-Health services. IEEE J. Biomed. Health Inform. **17**(1), 9–18 (2013). https://doi.org/10.1109/TITB.2012.2224876
14. Smith, E., Eloff, J.H.P.: Security in health-care information systems—Current trends. Elsevier Int. J. Med. Inform. **54**, 39–54 (2006)
15. Santamaria, N., Kapp, S.: TeleWound care—Providing remote wound assessment and treatment in the home care setting: current status and future directions. J. Smart Homecare Technol. TeleHealth **1**, 35–41 (2013)
16. Jones, V., Gay, V., Leijdekkers, P.: Body sensor networks for mobile health monitoring: experience in Europe and Australia. In: International Conference on Digital Society (ICDS), pp. 204–209. IEEE, St. Maarten (2010). https://doi.org/10.1109/ICDS.2010.41
17. Guo, L., Zhang, C., Sun, J., Fang, Y.: A privacy-preserving attribute-based authentication system for mobile health networks. IEEE Trans. Mob. Comput. **13**(9), 1927–1941 (2014). https://doi.org/10.1109/TMC.2013.84
18. Kumar, P., Lee, S.G., Lee, H.J.: E-SAP: efficient-strong authentication protocol for healthcare applications using wireless medical sensor networks. Sensors **12**(2), 1625–1647 (2012). https://doi.org/10.3390/s120201625
19. Navarro, K.F., Lawrence, E.: WSN applications in personal healthcare monitoring systems: a heterogeneous framework. In: International Conference on eHealth, Telemedicine, and Social Medicine, pp. 77–83, St. Maarten (2010). https://doi.org/10.1109/eTELEMED.2010.17
20. Kluge, E.H.W.: Secure e-health: managing risks to patient health data. Int. J. Med. Inform. **76**(5–6), 402–406 (2007). https://doi.org/10.1016/j.ijmedinf.2006.09.003
21. Abd Ghani, M.K., Wen, L.C.: The design of flexible pervasive electronic health record (PEHR). In: 2011 IEEE Colloquium on Humanities, Science and Engineering, CHUSER 2011, pp. 249–254 (2011). https://doi.org/10.1109/CHUSER.2011.6163727
22. Van Gorp, P., Comuzzi, M., Jahnen, A., Kaymak, U., Middleton, B.: An open platform for personal health record apps with platform-level privacy protection. Elsevier J. Comput. Biol. Med. 1–10 (2014). https://doi.org/10.1016/j.compbiomed.2014.04.019
23. Bai, G., Guo, Y.: Activity theory ontology for knowledge sharing in e-health. In: International Forum on Information Technology and Applications, pp. 39–43. IEEE, Kunming (2010). https://doi.org/10.1109/IFITA.2010.97
24. Nagaty, K.A.: Mobile health care on a secured hybrid cloud. J. Sel. Areas Health Inform. (JSHI) **4**(2) (2014)
25. Galarraga, M., Serrano, L., Martinez, I., de Toledo, P., Reynolds, M.: Telemonitoring systems interoperability challenge: an updated review of the applicability of ISO/IEEE 11073 standards for interoperability in telemonitoring. In: International Conference of the IEEE Engineering in Medicine and Biology Society, pp. 6161–6165. IEEE, Lyon (2007). https://doi.org/10.1109/IEMBS.2007.4353761
26. Yao, W., Chu, C.-H., Li, Z.: The adoption and implementation of RFID technologies in healthcare: a literature review. J. Med. Syst. **36**(6), 3507–3525 (2012). https://doi.org/10.1007/s10916-011-9789-8
27. Zhang, R., Liu, L.: Security models and requirements for healthcare application clouds. In: International Conference on Cloud Computing, pp. 268–275. IEEE, Miami (2010). https://doi.org/10.1109/CLOUD.2010.62
28. Kamarudin, N.H., Yussoff, Y.M., Hashim, H.: Two-tier e-health monitoring system. In: WSEAS Applied Computational Science (ACACOS), pp. 154–158, Kuala Lumpur (2015)
29. Jolly, R.: The e health revolution—Easier said than done. Dept of Parliamentary Services Library, no. 3 (2011)
30. Latré, B., Braem, B., Moerman, I., Blondia, C., Demeester, P.: A survey on wireless body area networks. Springer J. Wirel. Netw. **17**(1), 1–18 (2011). https://doi.org/10.1007/s11276-010-0252-4
31. Izzatty, N., Ismail, B., Hazana, N., Abdullah, B.: Developing electronic medical records (EMR) framework for Malaysia's Public Hospitals. In: IEEE Colloquium on Humanities, Science and Engineering Research (CHUSER), pp. 131–136. IEEE, Penang (2011). https://doi.org/10.1109/CHUSER.2011.6163702

32. Bahtiyar, S., Caglayan, M.U.: Trust assessment of security for e-health systems. Electron. Commer. Res. Appl. (2013). https://doi.org/10.1016/j.elerap.2013.10.003
33. Wang, F., Zhu, J., Shen, M.: A communication protocol of wireless sensor network for mobile healthcare system. In: International Conference on Green Computing and Communications & 2010 IEEE/ACM International Conference on Cyber Physical and Social Computing, pp. 554–559. IEEE, Hangzhou (2010). https://doi.org/10.1109/GreenCom-CPSCom.2010.98
34. Koutsouris, D., et al.: The use of telephone monitoring for diabetic patients: theory and practical implications. J. Smart Homecare Technol. TeleHealth **2**, 13–17 (2014). https://doi.org/10.2147/SHTT.S41242
35. Yeh, C.K., Chen, H.M., Lo, J.W.: An authentication protocol for ubiquitous health monitoring systems. J. Med. Biol. Eng. **33**(4), 415–419 (2013). https://doi.org/10.5405/jmbe.1478
36. Pantelopoulos, A., Bourbakis, N.G.: A survey on wearable sensor-based systems for health monitoring and prognosis. IEEE Trans. Syst. Man Cybern. Part C (Appl. Rev.) **40**(1), 1–12 (2010)
37. Moghaddam, F.F., Varnosfaderani, S.D., Ghavam, I., Mobedi, S.: A client-based user authentication and encryption algorithm for secure accessing to cloud servers based on modified Diffie-Hellman and RSA small-e. In: Proceeding—2013 IEEE Student Conference on Research and Development, SCOReD 2013, no. December, pp. 175–180 (2015). https://doi.org/10.1109/SCOReD.2013.7002566
38. Marcus, A., Cardei, I., Furht, B.: A mobile device prototype application for the detection and prediction of node faults in wireless sensor networks. J. Netw. Internet Arch. (2014)
39. Sun, Q., Song, W.: Thinking about some issues of e-health. J. US-China Public Adm. **10**(2), 209–216 (2013)
40. Singh, R., Singh, D.K., Kumar, L.: A review on security issues in wireless sensor network. J. Inf. Syst. Commun. **1**(1), 1–8 (2010)
41. Yussoff, Y.M., Hashim, H., Baba, M.D.: Identity-based trusted authentication in wireless sensor network. Int. J. Comput. Sci. Issues **9**(3) (2012)
42. Sen, J.: A survey on wireless sensor network security. Int. J. Commun. Netw. Inf. Secur. (IJCNIS) **1**(2), 59–83 (2009)
43. Dong, N., Jonker, H., Pang, J.: Challenges in eHealth: from enabling to enforcing privacy. In: International Conference on Foundations of Health Informatics Engineering and Systems (FHIES), pp. 195–206 (2012)
44. Li, M., Yu, S., Guttman, J.D., Lou, W., Ren, K.: Secure ad-hoc trust initialization and key management in wireless body area networks. J. ACM Trans. Sens. Netw.**9**(2) (2013). https://doi.org/10.1145/2422966.2422975
45. Singhal, D., Barjatiya, S., Ramamurthy, G.: A novel network architecture for cognitive wireless sensor network. In: 2011 International Conference on Signal Processing, Communication, Computing and Networking Technologies, no. Icsccn, pp. 76–80 (2011). https://doi.org/10.1109/ICSCCN.2011.6024518
46. Milenković, A., Otto, C., Jovanov, E.: Wireless sensor networks for personal health monitoring: issues and an implementation. Wirel. Comput. Commun. **29**(13–14), 2521–2533 (2006). https://doi.org/10.1016/j.comcom.2006.02.011
47. Hu, Y., Lu, F., Khan, I., Bai, G.: A cloud computing solution for sharing healthcare information. In: 7th International Conference for Internet Technology and Secured Transactions (ICITST), pp. 465–470, London (2012)
48. He, D., Zeadally, S.: An analysis of RFID authentication schemes for internet of things in healthcare environment using elliptic curve cryptography. IEEE Internet Things J. **2**(1), 72–83 (2015). https://doi.org/10.1109/JIOT.2014.2360121
49. Shen, P., Liu, V., Caelli, W.: A viable and sustainable key management approach for a national e-health environment. In: IEEE International Conference on e-Health Networking, Applications and Services, pp. 347–352 (2012)
50. Goncalves, F., Macedo, J., Nicolau, M.J., Santos, A.: Security architecture for mobile e-health applications in medication control. In: International Conference on Software, Telecommunications and Computer Networks, SoftCOM 2013, pp. 1–8, Primosten (2013). https://doi.org/10.1109/SoftCOM.2013.6671901

51. Khorshed, M.T., Ali, A.B.M.S., Wasimi, S.A.: A survey on gaps, threat remediation challenges and some thoughts for proactive attack detection in cloud computing. Elsevier Future Gener. Comput. Syst. **28**(6), 833–851 (2012). https://doi.org/10.1016/j.future.2012.01.006
52. Wilkowska, W., Ziefle, M.: Perception of privacy and security for acceptance of e-health technologies: exploratory analysis for diverse user groups. In: International Conference on Pervasive Computing Technologies for Healthcare (PervasiveHealth) and Workshops, pp. 593–600, Dublin (2011). https://doi.org/10.4108/icst.pervasivehealth.2011.246027
53. Lopes de Souza, R., Lung, L.C., Custodio, R.F.: Multi-factor authentication in key management systems. In: 12th IEEE international conference on trust, security and privacy in computing and communications, pp. 746–752. IEEE, Melbourne (2013). https://doi.org/10.1109/TrustCom.2013.90
54. Nguyen, H.A., Förster, A., Puccinelli, D., Giordano, S.: An experimental study of sensor node lifetime. In: International Workshop on Sensor Networks and Systems for Pervasive Computing (PerSens), pp. 202–207, Seattle (2011)
55. Rahim, Y.A., Sahib, S., Khanapi, M., Ghani, A.: Pseudonmization techniques for clinical data: privacy study in Sultan Ismail Hospital Johor Bahru. In: International Conference on Networked Computing (INC), pp. 74–77. IEEE, Gyeongsangbuk-do (2011)
56. Mahncke, R.J., Williams, P.A.H.: Secure transmission of shared electronic health records: a review. In: Proceedings of Australian information security management, Perth (2006)
57. Ghazvini, A., Shukur, Z.: Security challenges and success factors of electronic healthcare system. In: Elsevier International Conference on Electrical Engineering and Informatics, vol. 11, pp. 212–219 (2013). https://doi.org/10.1016/j.protcy.2013.12.183
58. Kiah, M.L.M., Nabi, M.S., Zaidan, B.B., Zaidan, A.A.: An enhanced security solution for electronic medical records based on AES hybrid technique with SOAP/XML and SHA-1. J. Med. Syst. **37**(5) (2013). https://doi.org/10.1007/s10916-013-9971-2
59. Choi, B., Kim, B., Lee, S., Wang, K., Kim, Y., Chung, D.: Narrowband physical layer design for WBAN system. In: International Conference on Pervasive Computing, Signal Processing and Applications, pp. 154–157. IEEE, Harbin (2010). https://doi.org/10.1109/PCSPA.2010.46
60. Huixian, L., Yang, Y., Pang, L.: An efficient authentication protocol with user anonymity for mobile networks. In: IEEE Wireless Communications and Networking Conference (WCNC), pp. 1842–1847. IEEE, Shanghai (2013). https://doi.org/10.1109/WCNC.2013.6554844
61. Kwak, K.S., Ullah, S., Ullah, N.: An overview of IEEE 802.15.6 standard. In: International Symposium on Applied Sciences in Biomedical and Communication Technologies (ISABEL), pp. 1–6. IEEE, Rome (2010). https://doi.org/10.1109/ISABEL.2010.5702867
62. Sood, S.K., Sarje, A.K., Singh, K.: A secure dynamic identity based authentication protocol for multi-server architecture. J. Netw. Comput. Appl. **34**(2), 609–618 (2011). https://doi.org/10.1016/j.jnca.2010.11.011
63. Fernando, J.I., Dawson, L.L.: The health information system security threat lifecycle: an informatics theory. Int. J. Med. Inform. **78**(12), 815–826 (2009). https://doi.org/10.1016/j.ijmedinf.2009.08.006
64. Chan, V., Ray, P., Parameswaran, N.: A multi-agent collaborative framework for mobile e-health. In: Proceedings of the Annual Hawaii International Conference on System Sciences, pp. 134–144, Waikoloa (2007). https://doi.org/10.1109/HICSS.2007.56
65. Othman, M.A., et al.: Wearable health monitoring system (WHMS) using wireless sensor network. Aust. J. Basic Appl. Sci. **7**(3), 93–105 (2013)
66. Yussoff, Y.M., Hashim, H.: Analysis of energy consumption on IBE-trust security framework. In: 17th Asia Pacific Conference on Communications (APCC), pp. 236–240, Kota Kinabalu (2011). https://doi.org/10.1109/APCC.2011.6152811
67. Javaid, N., Hayat, S., Shakir, M., Khan, M.A., Bouk, S.H., Khan, Z.A.: Energy efficient MAC protocols in wireless body area sensor networks—A survey. J. Basic Appl. Sci. Res. (JBASR) 1–17 (2013)
68. Gagnon, M.P., et al.: Systematic review of factors influencing the adoption of information and communication technologies by healthcare professionals. Springer J. Med. Syst. **36**(1), 241–277 (2012). https://doi.org/10.1007/s10916-010-9473-4

69. Bae, M., Lee, S.K., Yoo, S., Kim, H.: FASE: fast authentication system for e-health. In: 5th International Conference on Ubiquitous and Future Network (ICUFN), pp. 648–649. IEEE, Da Nang (2013). https://doi.org/10.1109/ICUFN.2013.6614898
70. Lee, C., Kim, J., Lee, H.S., Kim, J.: Physical layer designs for WBAN systems in IEEE 802.15.6 proposals. In: IEEE International Symposium on Communication and Information Technology (ISCIT), pp. 841–844. IEEE, Incheon (2009). https://doi.org/10.1109/ISCIT.2009.5341123
71. Pawar, P., Jones, V., van Beijnum, B.J.F., Hermens, H.: A framework for the comparison of mobile patient monitoring systems. J. Biomed. Inform. **45**(3), 544–556 (2012). https://doi.org/10.1016/j.jbi.2012.02.007
72. Chan, V., Ray, P., Parameswaran, N.: Mobile e-Health monitoring: an agent-based approach. IET J. Telemed. e-Health Commun. Syst. **2**(2), 223–230 (2008). https://doi.org/10.1049/iet-com
73. Thilakanathan, D., Chen, S., Nepal, S., Calvo, R., Alem, L.: A platform for secure monitoring and sharing of generic health data in the cloud. Elsevier Future Gener. Comput. Syst. **35**, 102–113 (2014). https://doi.org/10.1016/j.future.2013.09.011
74. Talaei-Khoei, A., Ray, P., Parameswaran, N.: An awareness framework for agent-based mobile health monitoring. In: International Conference on Next Generation Mobile Applications, Services and Technologies, pp. 108–113, Cardiff, Wales (2009). https://doi.org/10.1109/NGMAST.2009.39
75. Jafari, A.R., Encarnacao, A.Z., Dabiri, S.F., Noshadi Majid, H.: Wireless sensor networks for health monitoring. In: International Conference on Mobile and Ubiquitous System: networking and Services, pp. 479–481. IEEE (2005). https://doi.org/10.1109/MOBIQUITOUS.2005.65
76. Otoom, A.F., Abdallah, E.E., Kilani, Y., Kefaye, A., Ashour, M.: Effective diagnosis and monitoring of heart disease. Int. J. Softw. Eng. Appl. **9**(1), 143–156 (2015). https://doi.org/10.14257/ijseia.2015.9.1.12
77. Simplicio, M.A., Iwaya, L.H., Barros, B.M., Carvalho, T.C.M.B., Naslund, M.: SecourHealth: a delay-tolerant security framework for mobile health data collection. IEEE J. Biomed. Health Inform. **19**(2), 761–772 (2015). https://doi.org/10.1109/JBHI.2014.2320444
78. Yan, H.Y.H., Huo, H.H.H., Xu, Y.X.Y., Gidlund, M.: Wireless sensor network based e-health system : implementation and experimental results. IEEE Trans. Consum. Electron. **56**(4), 2288–2295 (2010). https://doi.org/10.1109/TCE.2010.5681102
79. Komnakos, D., Vouyioukas, D., Maglogiannis, I., Constantinou, P.: Feasibility study of a joint e-health mobile high-speed and wireless sensor system. In: Proceedings of the 1st ACM International Conference on PErvasive Technologies Related to Assistive Environments—PETRA '08, p. 1 (2008). https://doi.org/10.1145/1389586.1389615
80. Shnayder, V., Chen, B., Lorincz, K., Jones, T.R.F.F., Welsh, M.: Sensor networks for medical care. In: Proceedings of Third International Conference on Embedded Networked Sensor Systems, no. June, p. 314 (2005). https://doi.org/10.1145/1098918.1098979
81. Nagayama, T., Spencer Jr., B.F., Rice, J.A.: Structural health monitoring utilizing Intel's Imote2 wireless sensor platform. In: Sensors and Smart Structures Technologies for Civil, Mechanical, and Aerospace Systems 2007, Pts 1 and 2, vol. 6529, no. 1–2, pp. U1251–U1265 (2007). https://doi.org/10.1117/12.714629
82. O'Donovan, T., O'Donoghue, J., Sreenan, C., Sammon, D., O'Reilly, P., O'Connor, K.A.: A context aware wireless body area network (BAN). In: 2009 3rd International Conference on Pervasive Computing Technologies for Healthcare, pp. 1–8 (2009). https://doi.org/10.4108/ICST.PERVASIVEHEALTH2009.5987
83. Carnì, D.L., Fortino, G., Gravina, R., Grimaldi, D., Guerrieri, A., Lamonaca, F.: Continuous, real-time monitoring of assisted livings through wireless body sensor networks. In: Proceedings of the 6th IEEE International Conference on Intelligent Data Acquisition and Advanced Computing Systems: Technology and Applications, IDAACS'2011, vol. 2, no. September, pp. 872–877 (2011). https://doi.org/10.1109/IDAACS.2011.6072896

84. Garcia, J.J., Falck, T.: Quality of service for IEEE 802.15.4-based wireless body sensor networks. In: 3rd International Conference on Pervasive Computing Technologies for Healthcare, pp. 1–6 (2009). https://doi.org/10.4108/ICST.PERVASIVEHEALTH2009.6054
85. Yang, H., Kim, H., Mtonga, K.: An efficient privacy-preserving authentication scheme with adaptive key evolution in remote health monitoring system. Peer Peer Netw. Appl. **8**(6), 1059–1069 (2014). https://doi.org/10.1007/s12083-014-0299-6
86. Ullah, A., Aznaoui, H., Sebai, D., Abualigah, L., Alam, T., Chakir, A.: Internet of things and cloud convergence for e-health systems: concepts, opportunities, and challenges. Wirel. Pers. Commun. **133**(3), 1397–1447 (2023). https://doi.org/10.1007/s11277-023-10817-2
87. Martí, R.: Security in a wireless mobile health care system. In: International Conference on Emerging Application for Wireless and Mobile Access, vol. 8 (2005)
88. Debiao, H., Jianhua, C., Rui, Z.: A more secure authentication scheme for telecare medicine information systems. J. Med. Syst. **36**(3), 1989–1995 (2012). https://doi.org/10.1007/s10916-011-9658-5
89. Jiang, Q., Ma, J., Ma, Z., Li, G.: A privacy enhanced authentication scheme for telecare medical information systems. J. Med. Syst. **37**(1) (2013). https://doi.org/10.1007/s10916-012-9897-0
90. Bassham, L.E., Polk, T.: Threat assesment of malicious code and human threats. In: National Institute of Standards and Technology Computer Security Division. http://csrc.nist.gov/publications/nistir/threats/threats.html. Accessed 13 Jan. 2017
91. Ghazizadeh, E., Zamani, M., Ab Manan, J., Alizadeh, M.: Trusted computing strengthens cloud authentication. Sci. World J. (2014). https://doi.org/10.1155/2014/260187
92. Li, X., et al.: A novel chaotic maps-based user authentication and key agreement protocol for multi-server environments with provable security. Wirel. Pers. Commun. **89**(2), 569–597 (2016). https://doi.org/10.1007/s11277-016-3293-x
93. AbuKhousa, E., Mohamed, N., Al-Jaroodi, J.: E-health cloud: opportunities and challenges. Future Internet **4**(4), 621–645 (2012). https://doi.org/10.3390/fi4030621
94. Omar, W.M., Taleb-Bendiab, A.: E-health support services based on service-oriented architecture. IEEE J. Mag. IT Prof. **8**(2), 35–41 (2006). https://doi.org/10.1109/MITP.2006.32
95. Maitra, T., Giri, D.: An efficient biometric and password-based remote user authentication using smart card for telecare medical information systems in multi-server environment. J. Med. Syst. **38**(12) (2014). https://doi.org/10.1007/s10916-014-0142-x
96. Xu, L., Wu, F.: Cryptanalysis and improvement of a user authentication scheme preserving uniqueness and anonymity for connected health care. J. Med. Syst. **39**(2) (2015). https://doi.org/10.1007/s10916-014-0179-x
97. Chen, H.-M., Lo, J.-W., Yeh, C.-K.: An efficient and secure dynamic ID-based authentication scheme for telecare medical information systems. J. Med. Syst. **36**(6), 3907–3915 (2012). https://doi.org/10.1007/s10916-012-9862-y
98. Rahmani, A.M., et al.: Exploiting smart e-health gateways at the edge of healthcare internet-of-things: a fog computing approach. Future Gener. Comput. Syst. **78**, 641–658 (2018). https://doi.org/10.1016/j.future.2017.02.014
99. Harvey, M.J., Harvey, M.G.: Privacy and security issues for mobile health platforms. J. Assoc. Inf. Sci. Technol. **66**(7), 1305–1318 (2014). https://doi.org/10.1002/asi
100. Yüksel, B., Küpçü, A., Özkasap, Ö.: Research issues for privacy and security of electronic health services. Future Gener. Comput. Syst. **68**, 1–13 (2017). https://doi.org/10.1016/j.future.2016.08.011
101. Chatterjee, S., Das, A.K.: An effective ECC-based user access control scheme with attribute-based encryption for WSN. Secur. Commun. Netw. **9**(22), 5968–5974 (2014). https://doi.org/10.1002/sec.1140
102. Moosavi, S.R., et al.: SEA: a secure and efficient authentication and authorization architecture for IoT-based healthcare using smart gateways. Procedia Comput. Sci. **52**(1), 452–459 (2015). https://doi.org/10.1016/j.procs.2015.05.013

103. Chuang, M.-C., Chen, M.C.: An anonymous multi-server authenticated key agreement scheme based on trust computing using smart cards and biometrics. Expert Syst. Appl. **41**(4), 1411–1418 (2014). https://doi.org/10.1016/j.eswa.2013.08.040
104. He, D., Chen, J., Chen, Y.: A secure mutual authentication scheme for session initiation protocol using elliptic curve cryptography. Secur. Commun. Netw. (2012). https://doi.org/10.1002/sec.506
105. Wu, Z.Y., Chung, Y., Lai, F., Chen, T.S.: A password-based user authentication scheme for the integrated EPR information system. J. Med. Syst. **36**(2), 631–638 (2012). https://doi.org/10.1007/s10916-010-9527-7
106. Wu, Z.Y., Lee, Y.C., Lai, F., Lee, H.C., Chung, Y.: A secure authentication scheme for telecare medicine information systems. J. Med. Syst. **36**(3), 1529–1535 (2012). https://doi.org/10.1007/s10916-010-9614-9
107. Wei, J., Hu, X., Liu, W.: An improved authentication scheme for telecare medicine information systems. J. Med. Syst. **36**(6), 3597–3604 (2012). https://doi.org/10.1007/s10916-012-9835-1
108. Zhu, Z.: An efficient authentication scheme for telecare medicine information systems. J. Med. Syst. **36**(6), 3597–3604 (2012). https://doi.org/10.1007/s10916-012-9835-1
109. Khan, M.K., Kumari, S.: An authentication scheme for secure access to healthcare services. J. Med. Syst. **37**(4), 9954 (2013). https://doi.org/10.1007/s10916-013-9954-3
110. Islam, S.H., Khan, M.K.: Cryptanalysis and improvement of authentication and key agreement protocols for telecare medicine information systems. J. Med. Syst. **38**(10) (2014). https://doi.org/10.1007/s10916-014-0135-9
111. Lee, T.F., Liu, C.M.: A secure smart-card based authentication and key agreement scheme for telecare medicine information systems. J. Med. Syst. **37**(3) (2013). https://doi.org/10.1007/s10916-013-9933-8
112. Wang, Y., Liu, J., Xiao, F., Dan, J.: A more efficient and secure dynamic ID-based remote user authentication scheme. Comput. Commun. **32**(4), 583–585 (2009). https://doi.org/10.1016/j.comcom.2008.11.008
113. Khan, M.K., Kim, S.K., Alghathbar, K.: Cryptanalysis and security enhancement of a 'more efficient & secure dynamic ID-based remote user authentication scheme.' Comput. Commun. **34**(3), 305–309 (2011). https://doi.org/10.1016/j.comcom.2010.02.011
114. Xie, Q., Zhang, J., Dong, N.: Robust anonymous authentication scheme for telecare medical information systems. J. Med. Syst. **37**(2) (2013). https://doi.org/10.1007/s10916-012-9911-6
115. Wu, F., Xu, L.: Security analysis and improvement of a privacy authentication scheme for telecare medical information systems. J. Med. Syst. **37**(4), 1–9 (2012). https://doi.org/10.1007/s10916-013-9958-z
116. Arshad, H., Nikooghadam, M.: Three-factor anonymous authentication and key agreement scheme for telecare medicine information systems. J. Med. Syst. **38**(12) (2014). https://doi.org/10.1007/s10916-014-0136-8
117. Tan, Z.: A user anonymity preserving three-factor authentication scheme for telecare medicine information systems. J. Med. Syst. **38**(3) (2014). https://doi.org/10.1007/s10916-014-0016-2
118. Awasthi, A.K., Srivastava, K.: A biometric authentication scheme for telecare medicine information systems with nonce. J. Med. Syst. **37** (2013). https://doi.org/10.1007/s10916-013-9964-1
119. Das, A.K.: A secure user anonymity-preserving three-factor remote user authentication scheme for the telecare medicine information systems. J. Med. Syst. **39**(3) (2015). https://doi.org/10.1007/s10916-015-0218-2
120. Challa, S., et al.: An efficient ECC-based provably secure three-factor user authentication and key agreement protocol for wireless healthcare sensor networks. Comput. Electr. Eng. **0**, 1–21 (2017). https://doi.org/10.1016/j.compeleceng.2017.08.003
121. Fernández-Alemán, J.L., Señor, I.C., Lozoya, P.Á.O., Toval, A.: Security and privacy in electronic health records: a systematic literature review. J. Biomed. Inform. **46**(3), 541–562 (2013). https://doi.org/10.1016/j.jbi.2012.12.003
122. Diffie, W., Diffie, W., Hellman, M.E.: New directions in cryptography. IEEE Trans. Inf. Theory **22**(6), 644–654 (1976). https://doi.org/10.1109/TIT.1976.1055638

123. Shamir, A.: Identity-based cryptosystems and signature schemes. Adv. Cryptol. **196**, 47–53 (1985). https://doi.org/10.1007/3-540-39568-7_5
124. Boneh, D., Franklin, M.: Identity-based encryption from the Weil pairing. SIAM J. Comput. **32**(3), 586–615 (2003). https://doi.org/10.1137/S0097539701398521
125. Hölbl, M., Welzer, T.: Two improved two-party identity-based authenticated key agreement protocols. Comput. Stand. Interfaces **31**(6), 1056–1060 (2009). https://doi.org/10.1016/j.csi.2008.09.024
126. Huang, X., Craig, P., Wang, Q.: Identity-based association protocols for wireless personal area networks. In: International Conference on Computer and Information Technology; ubiquitous Computing and Communications; Dependable, Autonomic and Secure Computing; pervasive Intelligence and Computing, pp. 1819–1826 (2015). https://doi.org/10.1109/CIT/IUCC/DASC/PICOM.2015.272
127. Dill, D.L., Drexler, A.J., Hu, A.J., Yang, H.: Protocol verification as a hardware design aid. In: IEEE International Conference on Computer Design on VLSI in Computer & Processors (1992)
128. Holzmann, G.J.: The model checker SPIN. IEEE Trans. Softw. Eng. **23**(5), 279–295 (1997). https://doi.org/10.1109/32.588521
129. Larsen, K.G., Pettersson, P., Yi, W.: Uppaal in a nutshell. Int. J. Softw. Tools Technol. Transfer **1**(1–2), 134–152 (1997). https://doi.org/10.1007/s100090050010
130. Mitchell, J.C., Mitchell, M., Stern, U.: Automated analysis of cryptographic protocols using Murφ. Proceedings of the IEEE Symposium on Security and Privacy, pp. 141–151 (1997). https://doi.org/10.1109/SECPRI.1997.601329
131. Lowe, G.: Casper: a compiler for the analysis of security protocols. In: Proceedings of the Computer Security Foundations Workshop, pp. 18–30 (1997). https://doi.org/10.1109/CSFW.1997.596779
132. Blanchet, B.: An efficient cryptographic protocol verifier based on prolog rules. In: Proceedings. 14th IEEE Computer Security Foundations Workshop, pp. 82–96 (2001). https://doi.org/10.1109/CSFW.2001.930138
133. Armando, A., Basin, D., Boichut, Y., Chevalier, Y., Compagna, L.: The AVISPA tool for the automated validation. Comput. Aided Verif. **3576**, 281–285 (2005)
134. Chaudhry, S.A., Mahmood, K., Naqvi, H., Khan, M.K.: An improved and secure biometric authentication scheme for telecare medicine information systems based on elliptic curve cryptography. J. Med. Syst. **39**(11) (2015). https://doi.org/10.1007/s10916-015-0335-y
135. Chevalier, Y., et al.: A high level protocol specification language for industrial security-sensitive protocols. Austrian Comput. Soc. 350–357 (2004)
136. Dolev, D.: On the security of public key protocols. IEEE Trans. Inf. Theory **29**(2), 198–208 (1983). https://doi.org/10.1109/SFCS.1981.32
137. Das, A.K., Goswami, A.: An enhanced biometric authentication scheme for telecare medicine information systems with nonce using chaotic hash function. J. Med. Syst. **38**(6), 27 (2014). https://doi.org/10.1007/s10916-014-0027-z
138. Amin, R., Biswas, G.P.: A novel user authentication and key agreement protocol for accessing multi-medical server usable in TMIS. J. Med. Syst. **39**(3) (2015). https://doi.org/10.1007/s10916-015-0217-3
139. Tobarra, L., Cazorla, D., Cuartero, F.: Formal analysis of sensor network encryption protocol (SNEP). In: 2007 IEEE Internatonal Conference on Mobile Adhoc and Sensor Systems, MASS (2007). https://doi.org/10.1109/MOBHOC.2007.4428763
140. Tobarra, L., Cazorla, D., Cuartero, F., Díaz, G., Cambronero, E.: Model checking wireless sensor network security protocols: TinySec + LEAP + TinyPK. Telecommun. Syst. **40**(3–4), 91–99 (2009). https://doi.org/10.1007/s11235-008-9131-z
141. Odelu, V., Das, A.K., Goswami, A.: SEAP: Secure and efficient authentication protocol for NFC applications using pseudonyms. IEEE Trans. Consum. Electron. **62**(1), 30–38 (2016). https://doi.org/10.1109/TCE.2016.7448560

Trusted Access: Authentication Solutions for Mobile E-Health

Nazhatul Hafizah Kamarudin, Yusnani Mohd Yussoff, and Tutut Herawan

Abstract This chapter examines the development of a comprehensive passwordless authentication solution for mobile e-health systems, addressing challenges of security, data privacy, and user accessibility. As reliance on mobile health applications grows, especially for remote patient monitoring, traditional authentication methods using passwords are no longer sufficient, particularly for elderly users. The study introduces a multi-faceted approach, beginning with the creation of a secure mobile e-health application infrastructure, which eliminates the need for third-party servers. A unique identity-based encryption protocol is implemented, leveraging advanced cryptographic techniques to generate non-regenerated unique identities for e-health sensor nodes. This protocol is fortified through the integration of Near Field Communication (NFC) technology and secure boot processes, which enable trusted, device-specific authentication without relying on passwords. A mobile e-health test bed, combining hardware and software components, validates the feasibility of the authentication protocol under real-world conditions. The solution's efficacy is demonstrated through formal validation methods using tools like AVISPA and SPAN, showing resilience against node cloning and replay attacks. Furthermore, a lightweight identity-based encryption (IBE) algorithm is optimized to minimize memory usage and processing time, making it suitable for resource-constrained mobile devices. The findings reveal a significant enhancement in data security, privacy, and usability for mobile e-health applications. The proposed authentication framework provides a practical and scalable solution, benefiting elderly users by simplifying the authentication process and eliminating the need for password

N. H. Kamarudin (✉)
Center for Cyber Security, Faculty of Information Science and Technology, Universiti Kebangsaan Malaysia, Bangi, Malaysia
e-mail: nazhatulhafizah@ukm.edu.my

Y. Mohd Yussoff
School of Electrical Engineering, College of Engineering, Universiti Teknologi MARA, Shah Alam, Malaysia

T. Herawan
Department of Information Systems, Faculty of Computer Science and Information Technology, University of Malaya, Kuala Lumpur, Malaysia

N. H. Kamarudin et al. (eds.), *Beyond Passwords: Protecting Healthcare with Authentication Security*, Information Systems Engineering and Management 51,
https://doi.org/10.1007/978-3-031-92247-3_3

management. These contributions advance the field of mobile e-health, offering an innovative approach to secure authentication that aligns with the unique constraints of healthcare environments and supports a reliable, cost-effective infrastructure for future digital health systems.

Keywords Passwordless authentication · Mobile e-health security · Identity-based encryption (IBE) · Near field communication (NFC) · Network security

1 Introduction

This chapter delineates the comprehensive solution undertaken in this topic, aimed at crafting a robust passwordless authentication solution for mobile e-health systems. Comprising four pivotal subsections, this methodology offers a holistic approach to address the complexities inherent in securing mobile e-health environments. The chapter commences with an in-depth exploration of hardware and software development for the mobile e-health application. This section elucidates the intricacies involved in designing and implementing the requisite technological infrastructure to support secure authentication protocols. Subsequently, the focus shifts to the eradication of reliance on third-party servers within the mobile e-health network architecture. Through a systematic approach, this subsection delineates the process of transitioning towards a decentralized architecture, enhancing data privacy and system reliability. Following this, attention is turned to the technique employed in generating unique identities for e-health sensor nodes. By leveraging advanced cryptographic methodologies, this subsection outlines the procedures for generating non-regenerated identities, crucial for thwarting malicious cloning attacks.

Continuing the solution, detailed insights are provided into the development of a bespoke mobile e-health authentication protocol. This subsection elucidates the iterative process of protocol design, highlighting key considerations such as user experience, security, and interoperability. Moreover, the chapter expounds upon the rigorous methodology employed in the analysis of the designed authentication protocol. Through formal validation techniques and simulation studies, the efficacy and resilience of the protocol against various security threats are meticulously evaluated.

Lastly, the chapter culminates with a comprehensive summary, encapsulating the key findings and contributions of the mobile e-health solution. This synthesis provides a coherent overview of the steps undertaken and sets the stage for the subsequent chapters, offering insights into the development of a secure and user-friendly authentication framework for mobile e-health.

2 Hardware and Software Description

To begin with, mobile e-health test bed is developed in order to create a secure platform for the proposed authentication protocol. An e-health monitoring system is a significant implementation in mobile e-health and it needs to be developed meticulously in order to eliminate or to reduce security weaknesses which can be the target of the mobile e-health intruders. Mobile e-health test bed is developed through three core stages which are firstly, mobile e-health application development followed by identity-based encryption software development. Afterwards, near field communication ring application is developed to increase the security of the mobile e-health application. The mobile e-health test bed development can be referred in the picture shown in Fig. 1.

The implementation of the mobile e-health test bed is set up in the laboratory in which for the hardware implementation, a pulse sensor, an Arduino microcontroller, an Android mobile device and an in-laboratory server are utilized to create a secure platform for the mobile e-health authentication protocol implementation. The pulse sensor is connected to the Arduino microcontroller which will read the input data

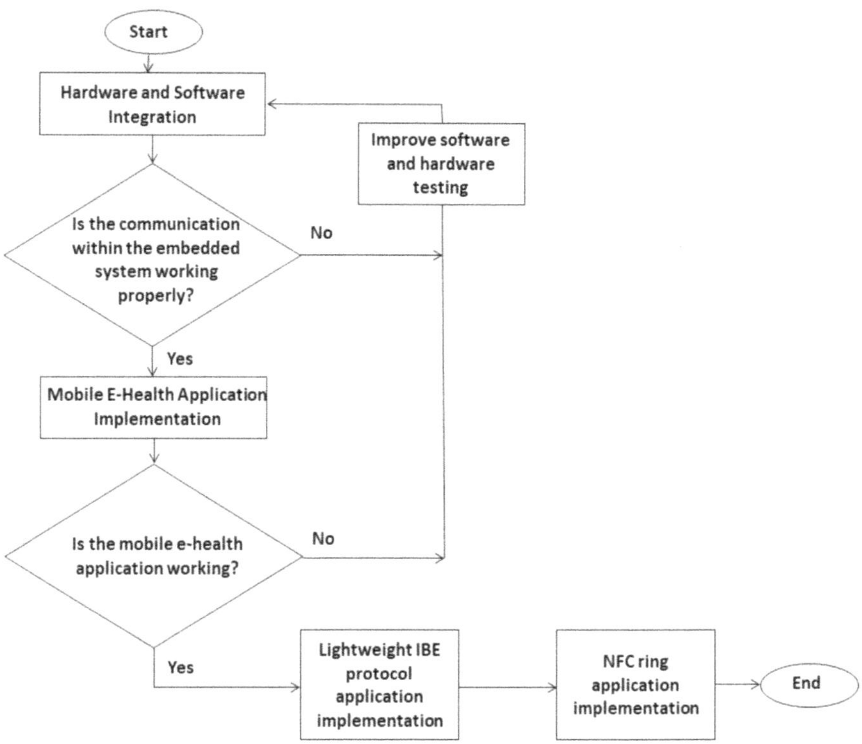

Fig. 1 Mobile e-health test bed development

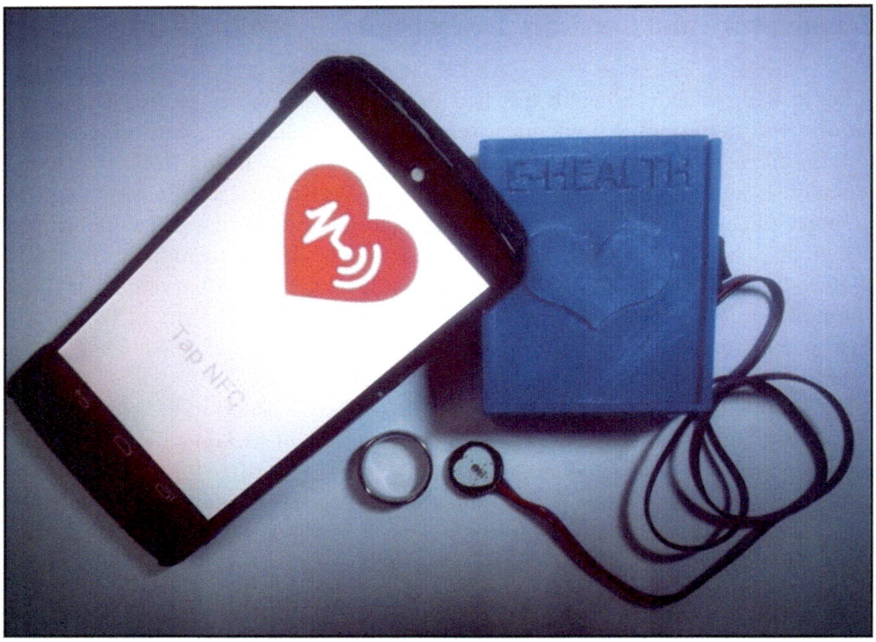

Fig. 2 Mobile e-health devices with the NFC ring implementation

and transfer the data to the mobile device through Bluetooth module HC-05. The front of the pulse sensor is the side with the heart logo where it is the side that can make contact with the user. This pulse sensor needs to be clipped to the fingertip of a user and connected to the embedded hardware.

Then, the data from the sensor node is transmitted to the mobile device through Bluetooth module and NFC ring implementation. Figure 2 shows the whole embedded mobile e-health device with the NFC ring implementation and mobile e-health application. NFC ring is used for the user verification as well as to turn on the Bluetooth module on the mobile device. Further details in each component will be explained thoroughly later in this chapter.

2.1 Mobile E-Health Application Development

Firstly, a secure platform for the mobile e-health system which is the mobile e-health application has to be developed in order to facilitate the security protocol that will be further implemented in the mobile e-health system. Three main activities are performed during the mobile e-health application development; hardware and software integration, mobile e-health application implementation, and mobile e-health application validation.

Table 1 Microcontroller specification

Microcontroller	ATmega328p @ 20 MHz
Operating voltage	5 V
I/O Pins	26
DC current	40 mA
Flash memory	32 KB
SRAM	2 KB
EEPROM	1 KB
Size (mm)	68.6 × 53.4

a. **Hardware and software integration**

The mobile e-health application is implemented on the Arduino computing platform with the hardware components consists of a pulse sensor and Bluetooth HC-05 module. This computing platform is developed with the microcontroller specifications shown in Table 1. HC-05 serial port Bluetooth is connected to the controller board in order to get the wireless data transmission while the pulse sensor reads the pulse rate value as an input value from the user to detect real time pulse data for e-health data storage as well as transmitting the sensor identity for user authentication. Bluetooth connection has to be successfully paired using Near Field Communication (NFC) ring that will be discussed in details in this later section.

The communication between the embedded system and the mobile phone is based on Bluetooth communication. Using Bluetooth HC-05 module, the e-health data is transmitted to the android mobile smartphone and the mobile e-health android application is developed focusing on the required specifications by using Arduino IDE 1.6.4 as well as Android Studio Developer. The e-health sensor node which is the pulse sensor is firstly connected to Arduino Uno and programmed using C programming language in order to give command to the microcontroller. The coding is then assembled and verified before being implemented into the Arduino microcontroller. Figure 3 shows the Arduino software programming tool that upload the coding instruction between the pulse sensor and Arduino microcontroller.

The mobile e-health system consists of a pulse sensor and a rechargeable battery which will act as an input mechanism and Bluetooth module HC-05 will be the output mechanism which will do the data transfer to the mobile device. Output LED will blink when the pulse sensor senses the BPM from the user (Fig. 4). The system microcontroller Atmega328p reads the input data from the pulse sensor and transmits the data via Bluetooth to the mobile e-health application. The pulse sensor needs to be clipped to the fingertip of a user and plugged into the embedded mobile e-health processor in order to get the pulse reading data.

A schematic layout design is the first step in the mobile e-health hardware development process. Atmega328p microcontroller is the brain of the mobile e-health application system that will do the software programming and further processing. An external 16 MHz crystal oscillator as well as 5 V battery supply is required in order to get the microcontroller to be working properly. Voltage regulator LM7805

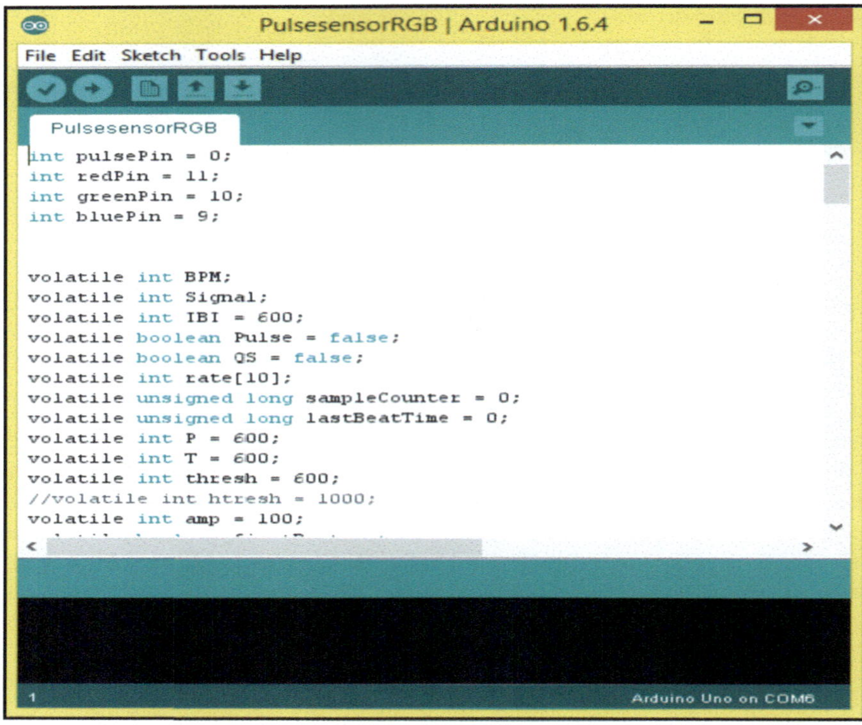

Fig. 3 Arduino software programming tool

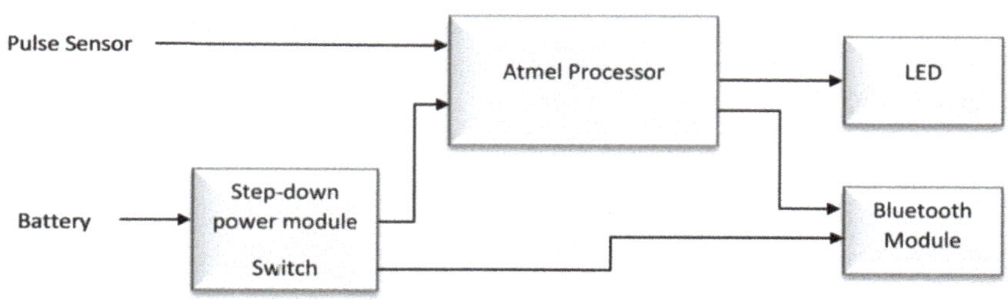

Fig. 4 Block diagram of the mobile e-health hardware development

is a 2 V linear drop-out and it must be power up to at least 7 V source to get a clean 5 V out. Constant current draw of 6mA is used in the system to produce a stable voltage for the microcontroller. Voltage regulator CJP1117 is used for voltage step down in order to get a 3.3 V power supply with a constant current draw of 5 mA. Figure 5 shows the details on the schematic layout design for mobile e-health hardware development.

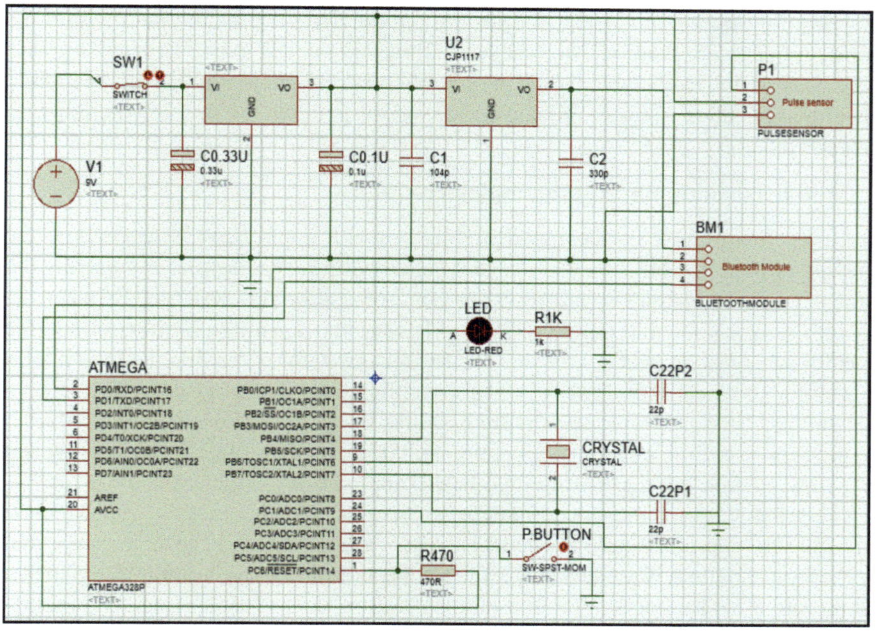

Fig. 5 Schematic layout design for mobile e-health hardware development

Next, the mobile e-health application is developed in the mobile device for better result display monitoring and it is designed to be e-health system user friendly especially for the elderly users. Processing Sketching software is used to develop data visualization of the e-health data waveform and to check the heart rate data from the pulse sensor node. Processing Sketching is the data visualization software which includes a text editor, a compiler, and a display window. The processing window is developed to show a graph of raw pulse sensor data over a period of time as well as a graph on beats per minute (BPM) that is updated every average last ten pulses. Android Studio Developer is an integrated development environment for android-based application and Java programming language is used. A Graphic User Interface (GUI) for mobile e-health application is designed and customized in order to display the value of the heart beat rate. The Android Studio software will generate an *.apk* application for the android-based smartphone to read and install it on the mobile device. Atmega328p can be programmed by using Arduino which provides an integrated development environment for the mobile e-health application system. Amarino platform is then installed in the mobile device in order to link the Arduino microcontroller with the mobile e-health application. Amarino is a toolkit that basically connects Android devices with Arduino microcontroller via Bluetooth. Amarino provides a platform that connects android-based mobile devices with Arduino microcontroller via Bluetooth module and it provides internal access to the mobile device that can be further processed using Arduino. Its application programming interface is used to transmit and receive the e-health data within the mobile e-health application. MeetAndroid library file needs to be extracted together with the Amarino executable jar file in order to integrate and to implement the mobile e-health application in the mobile device as well as to allow the user interaction with the Arduino.

b. Mobile e-health application implementation

The main function of the designed mobile e-health application is to become the gateway of the e-health sensor nodes and the e-health server besides monitoring the value of the e-health data waveforms. The communication protocol used between the e-health sensor node and mobile device is Bluetooth transmission protocol and the e-health sensor node needs to have a secure pairing with the mobile device initially to establish connectivity. Mobile e-health application is developed in XML file in the Android Studio Developer. Figure 6 shows the graphical user interface during the process of mobile e-health application and it shows the usage of ImageView and TextView method during the mobile e-health application development.

The graphical user interface (GUI) is implemented with LinearLayout which is Android GUI implementation. The required coding is written in Java language and the Android Studio will generate *.apk* file for the android-based mobile device so that the mobile e-health application can be installed on the mobile device. The mobile e-health application is developed to display the current pulse sensor in real time monitoring and RGB LED is installed to alert the user regarding the heart rate value condition (Fig. 7).

c. Mobile e-health application validation

The software validation needs to be done together with the hardware implementation. The whole coding is compiled into one full software program and the hardware testing is done to facilitate the mobile e-health requirement. The mobile e-health application

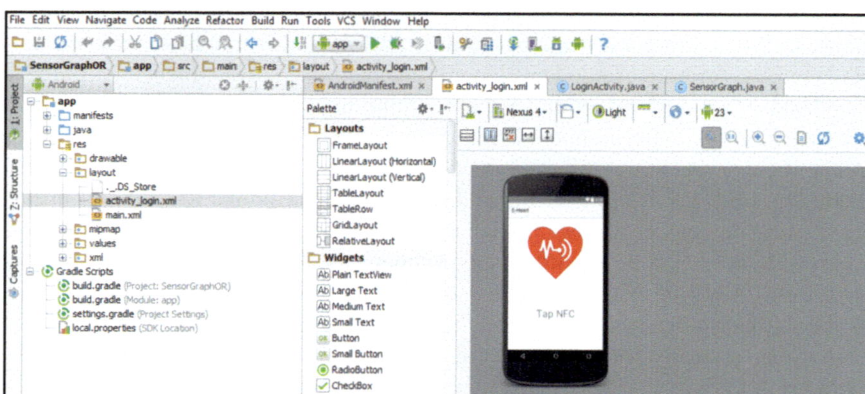

Fig. 6 Graphical user interface for mobile e-health application

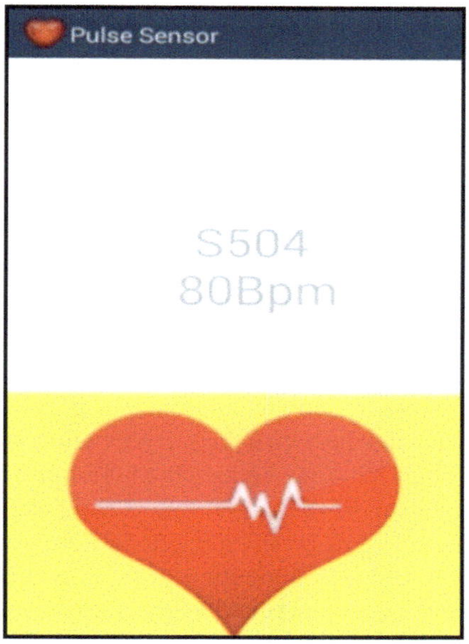

Fig. 7 Pulse sensor mobile e-health application

interface is firstly being implemented using the android emulator and Eclipse Luna application until the program is successfully working. Only then, the mobile interface is being implemented on a real mobile device and several troubleshooting are done in order to ensure that the mobile e-health application functions are in the right order. Troubleshoot on Bluetooth connection is done as well to validate the connection between the Arduino microcontroller and the pulse sensor node. In order to establish a more secure connection, the user needs to connect with only specified Bluetooth module and implement a designed near field communication (NFC) identity-based authentication in the mobile e-health application. Proposed NFC implementation will be fully discussed in later section.

2.2 *Identity-Based Encryption Protocol Implementation*

After mobile e-health application development, identity based authentication platform has to be implemented both in mobile device and e-health server. Lightweight identity based encryption application is developed in order to protect the e-health sensor node identity and passwordless mobile e-health authentication protocol will later be implemented in the mobile e-health application. The passwordless authentication protocol will be thoroughly discussed in Sect. 5. In this lightweight IBE scheme, an authentication code which contains the sensor node unique identity will be generated and Arduino microcontroller will be utilized as the trusted platform for the mobile e-health communication. Identity-based encryption is used in this network to validate the unique identity of the e-health sensor node, thus preventing the network from node impersonation or node cloning attack. IBE is a form of assy-metric cryptography based on elliptic curve cryptography (ECC) algorithms and it has smaller key size compared to RSA and AES. Before transmitting the unique identity of the e-health sensor node to the e-health server for further authentication process, it will first be encrypted using this proposed lightweight identity-based encryption protocol.

The normal IBE algorithm has basically four stages which are setup, extract, encrypt and decrypt. In the normal IBE algorithm, setup stage is needed to generate the master key while extract stage is implemented in order to extract the private key for the specified sensor node. Proposed lightweight IBE algorithm on the other hand combines the two earlier stages which are setup and extract into an only generated stage that performs the same function which generates the master key and the private key in only one single stage. The proposed generated stage is implemented in the lightweight IBE algorithm in order to be able to generate the master key as well as the private key which previously are generated in two stages during the normal IBE algorithm. Combination of these two IBE stages is able to be done since e-health server is programmed to generate the master key and extract the private key for each e-health sensor node during the registration phase in the mobile e-health authentication protocol which will be discussed further in Chap. "Protocols for Mobile E-Health Authentication". Protocol efficiency analysis and the comparison analysis are done to analyse the proposed generated phase. Details on the protocol efficiency analysis of the proposed lightweight IBE are presented in Table 8 and Fig. 5 in Chap. "Performance Analysis of Mobile E-Health Authentication System".

Next, the proposed lightweight IBE algorithm focus on minimizing the encrypt and decrypt algorithms. The overall development used Tate pairing algorithm and it has significantly reduce the memory usage at the e-health server up to sixty-five percent compared to the normal IBE algorithm. Details on the memory utilization analysis of the proposed lightweight IBE algorithm can be referred in Table 7 and Fig. 4 in Chap. "Performance Analysis of Mobile E-Health Authentication System". The overall process for the IBE algorithm minimization is shown in Fig. 8.

This lightweight IBE algorithm focuses on minimizing computational requirements and producing lower storage consumption in order to facilitate the mobile device constraints plus it is suitable to be implemented in mobile e-health application. The proposed lightweight encryption process does not require any communication from a third party or certificate generator which on the other hand is needed in a certificate based public key scheme. Only the recipient public ID is required to encrypt the login information. Arduino Yun is used as the trusted microcontroller to generate the IBE application since it is based on Atheros AR9331 processor that can support Linux distribution which is OpenWrt-Yun. Table 2 shows the specifications of Arduino Yun microcontroller. This IBE scheme focuses on the encryption and decryption processes of the sensor node unique identity and minimizing the IBE algorithms in order to produce lower processing time and less power consumption.

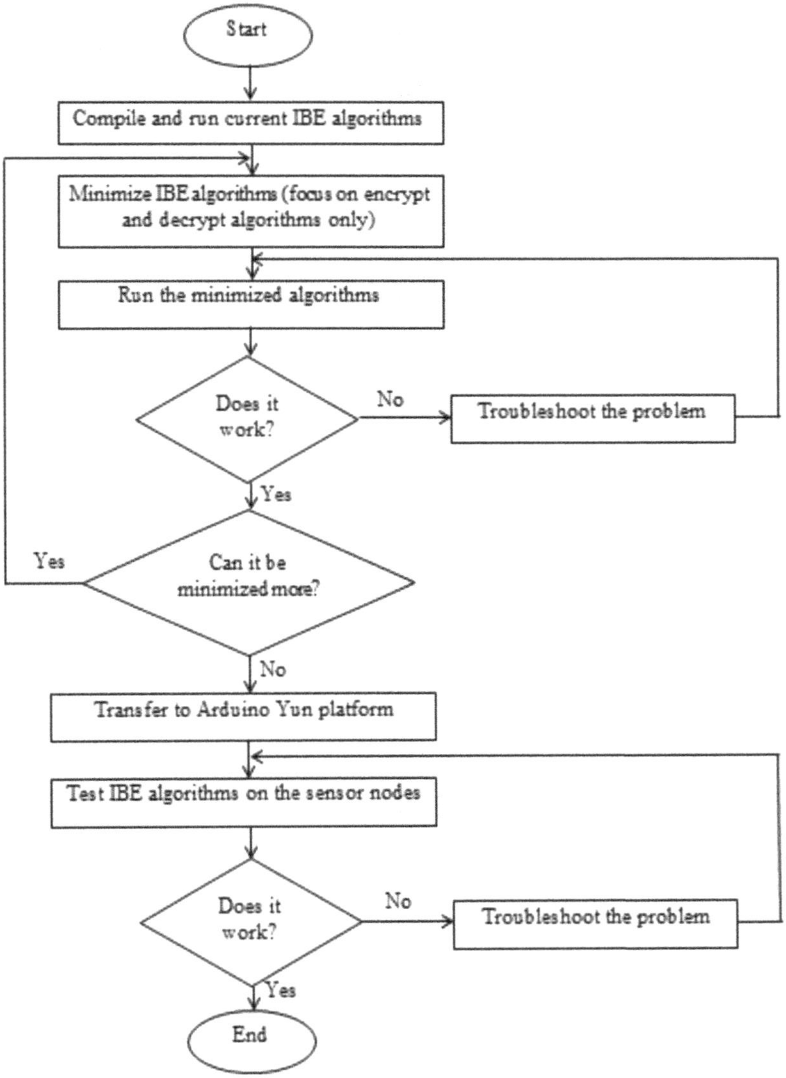

Fig. 8 IBE algorithm minimization process

Since mobile e-health and Arduino platform have a very limited constraint to implement complex encryption cryptography, this proposed lightweight IBE application also focuses on reducing memory utilization in order to improve device feasibility.

Firstly, current Stanford IBE library of Boneh-Franklin algorithms that was developed under Linux distribution is compiled and the source code is named *ibe-0.7.2.tgz*. Several libraries such as GNU Multiple Precision Arithmetic (GMP) and OpenSSL

Table 2 Arduino Yun specification

Microcoprocessor	Atheros AR9331
Operating voltage	3.3 V
SRAM/RAM	64 MB
Flash memory	16 MB

are required to compile the IBE coding. Then, this proposed lightweight IBE algorithm uses a bootable Knoppix Flash Drive to boot and run Linux distribution. It is required to open the terminal and run it as root in order to be able to modify the IBE files. Then, a *makefile* is needed to mechanically build and manage all the source files for this proposed lightweight IBE application. Few modifications need to be fixed since the current IBE algorithms occupy a large size of memory requirements hence it is not suitable to be implemented in the mobile e-health application. The most important procedure in this IBE algorithm development is to minimize the current memory size by analysing each detail of the source coding files. The following source files for the proposed lightweight IBE algorithm are presented in Table 3 with the description of each procedure needed. Only thirteen source files are developed in the proposed lightweight IBE algorithm to facilitate the limited constraint in mobile e-health compared to 36 source files in the current IBE algorithm which is not suitable to be implemented in mobile e-health application.

Table 3 IBE algorithm source files

Source files	Description
fp2.c	Performs a large and precise arithmetic procedure
curve.c	Executes the elliptic curve arithmetic that defines the points on the curve. Only specific curves are supported to apply certain optimizations
ibe_lib.c	Performs basic identity based encryption operations by defining the security parameters
crypto.c	Provides encryption, decryption and random number generation process
format.c	Possesses a security coding to save and load various file formats
byte_string.c	Defines a data type that holds a string of bytes
gen.c	Generate system parameters and split the master secret key
combine.c	Combines all private key shares and generate only a private key
encrypt.c	Contains a coding to encrypt the login information with a known ID
decrypt.c	Contains a coding to decrypt the login information with a given private key for that particular ID
ibe.c	A command-line program that performs the expected operations
ibe.h	An application programmer that builds lightweight IBE application and has the access to encryption and decryption library
*mm.** *benchmark.**	For debugging and benchmarking purposes

Lightweight IBE algorithm is developed by reducing the memory size of the system to increase feasibility of the mobile e-health application. After developing the lightweight IBE algorithm, it is then transferred to the Arduino Yun microcontroller in order for the user to have an access to a command line through Arduino library. During the encryption process, the output generated is the ciphertext with the secret keys. In the decryption process, the system is trying to decrypt those ciphertext together with the secret keys. Figure 9 shows the encryption and decryption process using the proposed lightweight IBE algorithm.

Lightweight IBE application is then installed in the mobile device in order to encrypt the unique identity of the e-health sensor node before transmitting it to the e-health server. It is very important to encrypt the unique identity of the e-health sensor node as well as other login information that might be used throughout the authentication process in order to minimize the security breaks and to enhance the security system itself. Figure 10 shows the lightweight IBE application in the mobile device and the unique identity that should be encrypted before transmitting over the mobile e-health channel.

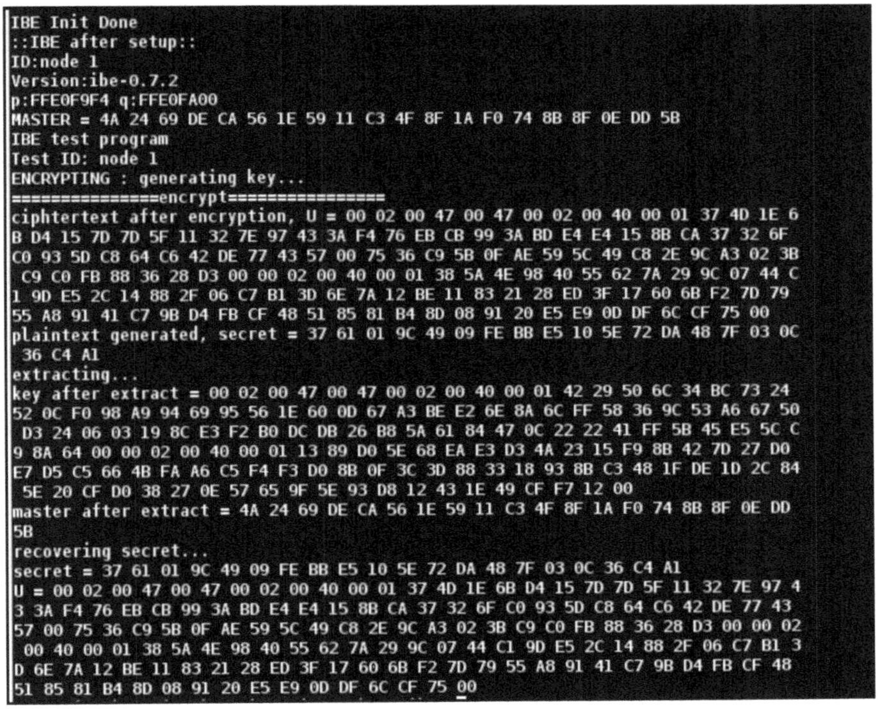

Fig. 9 Encryption and decryption process in lightweight IBE application

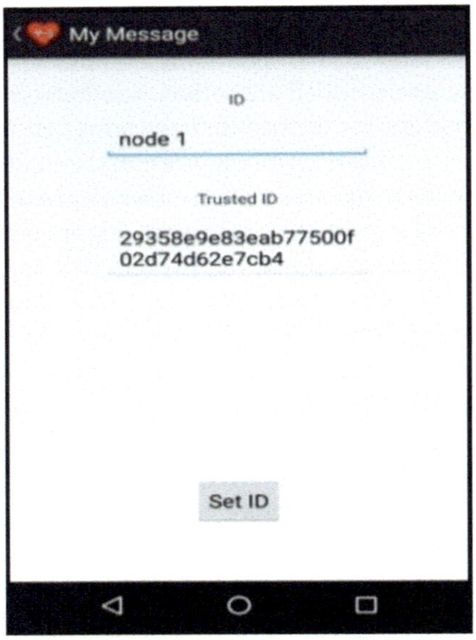

Fig. 10 Lightweight IBE application

2.3 Near Field Communication (NFC) Ring Application

Mobile e-health application is developed using Arduino microcontroller and the e-health sensor node is connected to the mobile device through Bluetooth module communication. In order to enhance the communication system between the e-health sensor node and the mobile device, NFC ring application is implemented in this mobile e-health to secure the Bluetooth module communication. The e-health sensor node will be connected to mobile device using Bluetooth communication using the implementation of Atmega328p microcontroller in the mobile e-health channel. Identity verification development of the NFC ring is developed to validate the communication between the e-health sensor node and the mobile device. The mobile device used in this mobile e-health application must support NFC technology and it is applied on the Android based operating system.

Secure device pairing concept is implemented in this section to protect the communication between the embedded e-health system and mobile device android application which is based on Bluetooth module. The e-health sensor node will transmit the data as well as its unique identity to the mobile device using Bluetooth channel. The lightweight IBE algorithm has been developed in order to encrypt the data during the transmission process towards the mobile device application. However, extra security layer has to be developed within the communication channel and this proposed NFC ring validation is implemented in order to successfully turn on the mobile e-health application. Atmega328p microcontroller requires an external 16MHz crystal oscillator and 5 V power supply in order to create electrical signal with precise frequency for NFC ring identity verification. Voltage regulators LM7805 and CJP1117 are used

Table 4 Bluetooth module specification

Standard	Bluetooth HC-05
Modulation	3 Mbps
ISM	2.4 GHz
Flash	8 MB
Version	V2.0 + EDR

Table 5 NFC ring specifications

Type	ISO.IEC 14443 (Type A)
Tag	Mifare ultralight C
Manufacturer	NXP semiconductors Germany

to get the exact power supply using a step down power module. Bluetooth module HC-05 is used as the communication channel and it is connected to the Atmega328p microcontroller.

Table 4 shows the specification of the Bluetooth module used in this e-health data transmission.

The NFC ring is an electronic ring that will be implemented during the sensor node authentication process with the mobile device. Only an authorized NFC ring with a valid e-health sensor node is able to turn on the Bluetooth module and get through into the mobile e-health application. NFC ring user interface is then developed in the mobile device and the user needs to touch the NFC ring with the mobile device for authentication process. Table 5 shows the specifications for the NFC ring that is used during this application development.

In order to perform NFC ring identity verification, the user needs to turn on the e-health sensor node and open the mobile e-health application first. Then, the user has to touch their NFC ring on the NFC spot area at the back of the mobile phone for identity verification. Only after a successful verification of the NFC ring identity, then the Bluetooth module will be activated and a secure device pairing connection between the e-health sensor node and the mobile device is created (Fig. 11).

There are basically two phase of authentication implemented during the mobile e-health communication which are the NFC ring identity verification and the e-health sensor node unique identity verification which will be discussed further in this chapter. Figure 12 shows the NFC ring user interface in the mobile e-health application.

This security implementation is developed to enhance the security of the mobile e-health application system and to reduce the mobile battery usage in Bluetooth communication. The Bluetooth connection will be turned off simultaneously with the mobile e-health application. Therefore, besides improving the security layer in mobile e-health, NFC ring implementation is developed to create a more reliable application in mobile e-health system.

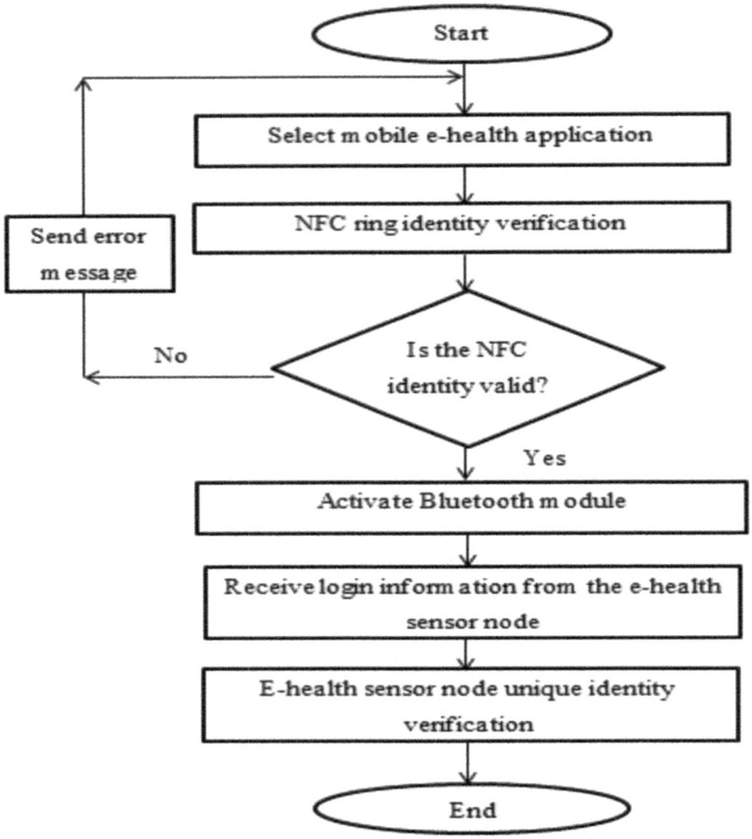

Fig. 11 NFC ring verification process

3 Eliminating the Third Party Server in the E-Health Network Architecture

Healthcare organizations and hospitals are really in need to implement their own e-health server so that they can eliminate the function of e-health service provider as well as cloud service provider. In order to improve the security of mobile e-health as well as reducing the high cost in mobile e-health implementation, e-health server needs to be installed in the hospital or medical centre. Three-tier mobile e-health network architecture implements a third party server in the network to generate the security parameters for their authentication protocols. By doing so, complex cryptography that requires a third party server such as certificate authority and PKG is not suitable to be implemented since the hospitals will need to include them in the mobile e-health security architecture. Thus, unique identity based technique for

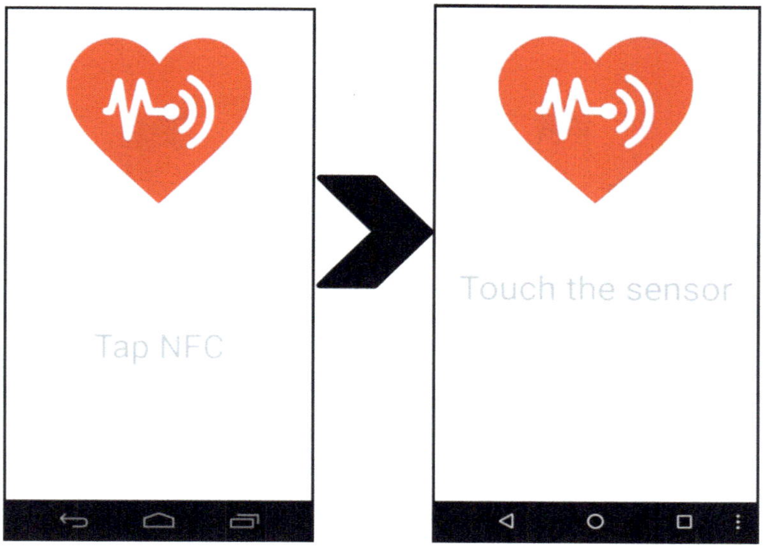

Fig. 12 NFC ring user interface

authentication scheme is utilized in the proposed authentication security protocol in order to eliminate the requirement for certificate verification and security parameters generation.

Development of a unique identity of the e-health sensor node has eliminated the function of a third party server in the current three-tier mobile e-health network architecture. The proposed authentication protocol does no longer need to go through a third party server since the identity of a user which is the e-health sensor node and the identity of the e-health server are the main entities needed for the security parameters generation. Previously, three tier mobile e-health network implements e-health sensor node, mobile device, e-health service provider server, and hospital server in the system. The data from the sensor will be transmitted over the channel through mobile device as a gateway node and will be sent to a third party server first for either verification process or data storage. Then only, the data will be transmitted to the hospital server. Figure 13 shows the three-tier communication from the e-health sensor node to the e-health server.

Identity-based encryption (IBE) implements public key cryptography and the usage of the e-health sensor node unique identity will eliminate the need for certificate authority or a third party server in the e-health network architecture while securing the e-health data. By implementing identity based encryption in the e-health authentication protocol, the need for certificate authority in the certificate based scheme is no longer needed plus it requires large memory utilization and higher computing time for encryption and decryption processes. Lightweight IBE algorithm as well as proposed passwordless mobile e-health authentication protocol is implemented to eradicate the requirement for a third party server. Moreover, lightweight IBE security

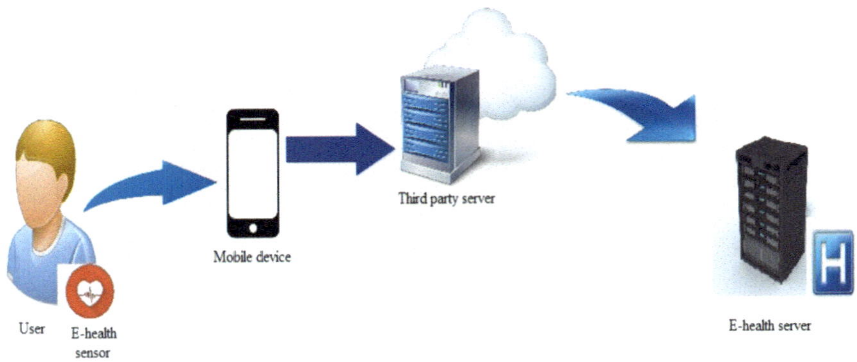

Fig. 13 Three tier mobile e-health network architecture

algorithm has minimal memory usage and power consumption. Thus, to minimize the communication requirements and to eliminate the function of a third party server in mobile e-health network architecture, identity-based authentication as well as identity based encryption has to be implemented in the proposed mobile e-health authentication framework.

4 Development of a Non-regenerated Unique Identity for E-Health Sensor

With the rise of security issues in mobile e-health system, it is crucial to have more than just a username and password safeguarding the e-health data. Username and password scheme is no longer a secure authentication for the mobile e-health application due to its highly sensitive and privacy element which dealing with medical information data. A unique and non-regenerated identity is developed using the e-health sensor node itself instead of using the biometric identity of a user to eliminate the risk of biometric pattern varying and denial of services attack during the authentication process. This section discusses and presents the development process of a non-regenerated unique identity for the e-health sensor node using a secure boot process as well as chain-identity-attestation (CIA) method. E-health sensor node is intensely constrained by power limitation and computation capability due to the smaller physical size of the e-health sensor node itself compared to other embedded system. Meticulous steps are required for security algorithm development to produce a unique identity for mobile e-health authentication protocol implementation.

4.1 Secure Boot Process

There are several methods for user identifications in WSN however the most reliable and secure method is by extracting the serial number in a microprocessor or a microchip to identify a sensor or a user [1]. Property of a sensor node platform is unique and infrangible. Sensor node platform chipset is one of the properties that will never change once it has been deployed. Therefore, integrity measurement process is done to obtain the identity of the sensor node platform which is securely located in Platform Configuration Register (PCR). The core root of trust measurement (CRTM) value is the first entity value that resides in the basic-input-output-system (BIOS) and that value is used to measure the next component in the sensor node platform. In this process, a newly joined e-health sensor node will boot up and go to a secure boot process which is the integrity measurement process in order to obtain the unique identity property located in PCR. With the purpose of obtaining that unique identity for the e-health sensor node, the integrity measurement process is done to locate the CRTM value in the on-chip sensor node microprocessor. The CRTM value is assumed trusted and cannot be modified. Therefore the integrity of the first layer of secure boot process is assumed to be always true. Then, the entity of the second layer residing in the external storage of the sensor node is verified by measuring the hash value of the entity. The secure boot process for the e-health sensor node is completed only if the integrity of the second entity value is verified. The secure boot process utilizes Secure Hash Algorithm (SHA) 256 implementation executed on the e-health sensor node. In order to minimize the register space in the e-health sensor node platform, the secure boot design considers only eight hexadecimal characters for identity validation. Figure 14 shows the components of the platform module which the CRTM value is located.

The secure boot process [2] is done to locate the unique property of the e-health sensor node in the PCR which is then applied as the e-health sensor node unique identity. It also captures the e-health sensor node serial number in order to generate

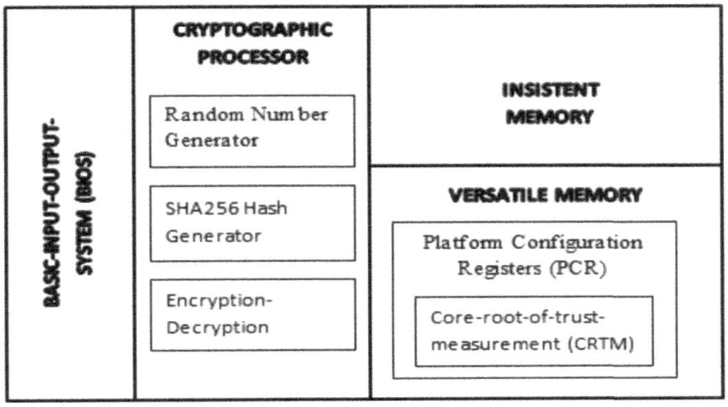

Fig. 14 Sensor node platform module

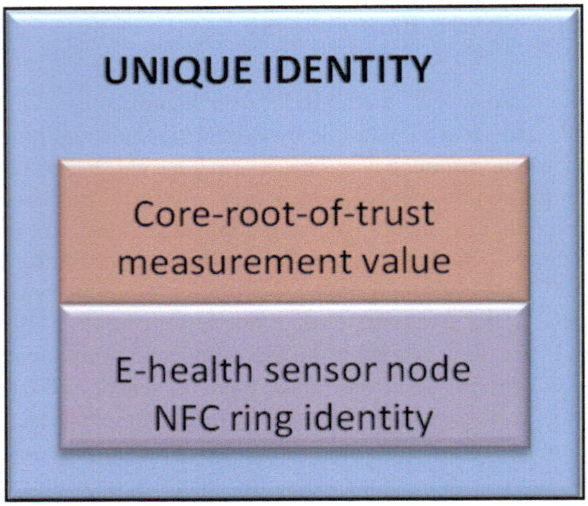

Fig. 15 E-health sensor node unique identity components

the unique identity for the e-health sensor node. Hence, CRTM value will be hashed together with the e-health sensor node NFC ring identity using SHA256 algorithm to produce that unique identity. Figure 15 shows the components of the unique identity of the e-health sensor node. SHA512 is not implemented in the research even though it offers higher security lever compared to SHA256 because it utilizes large memory usage which is not suitable for the lightweight mobile e-health scheme.

5 Implementation of the E-Health Authentication Protocol

Complex and complicated security algorithms are not suitable to be implemented in mobile e-health system due to limited constraint of the e-health devices. Limited constraints on the mobile e-health system have been discussed extensively in Chap. 2. Upon successful validation on the lightweight IBE algorithm and generating the unique identity for the e-health sensor node, this research proceeds in designing the e-health authentication protocol for the mobile e-health system. There is an urgent demand to secure the mobile e-health application domain in order to prevent fraudulent activities during the e-health data communication. The proposed authentication scheme for mobile e-health consists of a wireless body area network with an e-health sensor node, a personal mobile device and an e-health server. Current mobile e-health security mechanism is not clearly designated in a way to secure the e-health data and to facilitate the mobile e-health restricted constraints.

The designed EHEART authentication protocol acts as the final step in the development of a secure mobile e-health framework. It also ensures secure communications between the e-health sensor nodes and the e-health server. There are four important phases in the authentication protocol development which are registration phase, login phase, authentication phase, and also sensor-mobile change phase. Details on the designed protocol will be discussed in depth in Chap. 4. The designed protocol eliminates the need for username and password scheme and offers a seamless authentication in mobile e-health. In this final stage of developing the mobile e-health authentication framework, the designed authentication protocol provides a secure platform to verify the evidence from the remote spot which is the e-health sensor node. The e-health sensor node will report its unique identity that is generated during the secure boot process to the mobile device. The unique identity is encrypted and it is transmitted as a session identity to the mobile device through CIA model development. Using lightweight IBE implementation, the session identity of the e-health sensor node is then encrypted before it is transmitted to the e-health server for further attestation process. The e-health server will then verify the identity of the e-health sensor node based on the received entity compared with the pre-installed value in its database.

This research decides to use identity based public key cryptography system in authenticating the packet received at the remote platform which is the data from the e-health sensor node. ECC algorithm as well as Diffie-Hellman protocol is used in the identity based encryption implementation. There are basically four basic phases in the current IBE algorithm which are setup, extract, encrypt, and decrypt. During the setup phase, the IBE algorithm creates a master key for the private key generator (PKG) and using the master key, the IBE algorithm creates a private key during the extract phase. Encrypt phase is executed by the sender which is the mobile device and it needs to know and specify the receiver's identity which is the e-health server in order to transmit the e-health data. Last but not least, decrypt phase will be executed by the e-health server that possesses the private key generated during the extract phase which will be used to decrypt the encrypted data. Details steps involved in the identity based cryptography algorithm will be discussed together with the framework of mobile e-health authentication in Chapter Four. This research concentrates on the design of the mobile e-health authentication protocol and verifies its functionality by means of formal analysis. Figure 16 presents the overall list of basic building blocks toward a secure mobile e-health authentication platform.

By implementing ECC algorithm instead of RSA [3], the same level of security can be achieved with smaller key size in the proposed EHEART authentication protocol. 160-bit ECC gives the same security level as 1024-bit RSA and only 224-bit ECC provides the same comparable security as 2048-bit RSA [4]. Since the packet size and number of packets in data transmission are the important factors in the authentication protocol, ECC algorithm instead of RSA is chosen to be implemented in the proposed protocol together with Diffie-Hellman algorithm in order to gain more energy and computational efficiency.

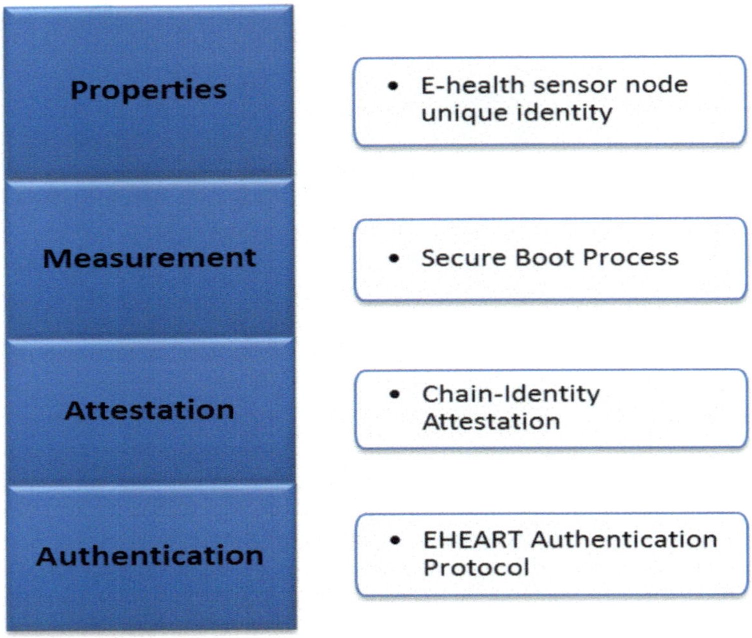

Fig. 16 Basic building block of a secure mobile e-health authentication platform

6 Analysis on the E-Health Authentication Protocol

Finally, to confirm and to verify the functionality and feasibility of the proposed mobile e-health authentication protocol as well as the security of the proposed protocol, performance analysis, power analysis and formal analysis on the protocol are structured. Following subsection present the methods used in analysing the authentication protocol.

6.1 Energy Analysis

To verify the practicability of the proposed mobile e-health authentication protocol, this research calculates energy consumption for a newly joined e-health sensor node during its unique identity registration with the e-health base station. The e-health sensor node platform processor runs at 20 mA, 3.6 V with the frequency of 667 MHz. The energy equation used and the running time of each process is measured by executing the mobile e-health application. Since encryption and communication processes utilize most of the total energy and power consumption [5], this study only considers the amount of energy used in the process. Hence, total energy for sending the encrypted data which is the unique identity of the e-health sensor node to the e-health server is calculated as in the Eq. (2).

$$E_{total} = E_{boot} + E_{authenticate} + E_{enc/bit} * data(bits) \tag{2}$$

$$E_{authenticate} = E_{TX} * bytes\ transmitted + E_{RX} * bytes\ received \tag{3}$$

where the rate of energy transfer is:

$$Power(Watts) = E(Joules)/time(seconds) \tag{4}$$

6.2 *Performance Analysis*

Performance analysis is done by conducting the memory analysis as well as protocol efficiency analysis. Memory analysis is performed to ensure that the proposed authentication protocol does not burden the mobile e-health application with heavy data and tons of software images. To generate details on the memory information, e-health sensor node compiler instruction command is used in the coding compilation to provide the total images and data size for each object file. Besides, this study also calculates each data or information that needs to be stored throughout the e-health sensor node in the e-health network to confirm the lightweight criteria of the proposed protocol. The data includes the e-health sensor node unique identity and number of keys. Protocol efficiency analysis is also conducted to compare the efficiency of the proposed protocol with previous implementation on authentication protocols and session key establishment. The protocol efficiency is based on the selected high computation process such as multiplication, hash and pairing that occur in the cryptographic algorithms.

6.3 *Formal Security Analysis*

Finally, the formal analysis method that is used to analyse and to validate the proposed authentication protocol for mobile e-health is presented in this subsection. As mention before, model checking tool AVISPA and Security Protocol Animator are used to finalize the analysis on the mobile e-health authentication protocol and to support the security analysis. The proposed communication events in the mobile e-health authentication protocol are thoroughly verified on its functionality. AVISPA uses a role-based language that involves basic roles and composed roles. A basic role defines initial knowledge as well as initial state and it also defines a set of transitions that describes the behaviour of the protocol participant. Multiple basic roles are combined together to form a composed role in which each protocol participant will have its basic roles specifying its sequence of actions. Figure 17 shows the basic role of the e-health sensor node while Fig. 18 shows a composed role between the e-health sensor node and mobile device.

The developed protocol model is converted into an intermediate format (IF) and selected to any of the four backends which are SATMC, OFMC, Cl-Atse and TA4SP

```
role sensor (SN, M        : agent,
         IDSN,IDM         :text,
         Mac,Hash         : hash_func,
         Ksn              :public_key,
         Snd,Rcv :channel(dy))

played_by SN def=

local State : nat,
      Nsn   : text,
      X     : text,        %random number
      W     : hash(text.public_key.text.text),    %H(X1.Ksn.IDSN.IDM) =
symmetric key
      X1    : text.public_key,   %X1=X.Ksn  to be send to mobile device M
      Ksnm     : symmetric_key,    % e((X1+W)Ssn,Km) where Sa=inv(Ksn)
      Km    :public_key
```

Fig. 17 Basic role of the e-health sensor node

```
role session (SN,M:agent,
              IDSN,IDM : text,
              Km,Ksn: public_key,
             Hash,Mac:hash_func)


def=

local SSN,RSN,SM,RM: channel (dy),
      BS      :(text) set,
         IDI : text
init BS :={IDSN,IDM}


composition

sensor(SN,M,IDSN,IDM,Mac,Hash,Ksn,SSN,RSN)
/\mobile(SN,M,IDSN,IDM,Hash,Mac,BS,SM,RM)


end role
```

Fig. 18 Composed role of the e-health sensor node and mobile device

for protocol verification. Each backend uses different techniques and balances each other. Dolev-Yao intruder model [6] is implemented in AVISPA to validate and to analyse the authentication security protocols. By implementing Dolev-Yao intruder model in the system, it is able to eavesdrop and intercept the message as well as modify and inject new message during the communication channel. However, Dolev-Yao intruder model is unable to compromise the end system application. SAT-based Model-Checking (SATMC) tool automatically translates the authentication security protocol specifications into propositional logic that is used to find the security attacks of the designed protocol.

On-The-Fly-Model-Checker (OFMC) tool [7] on the other hand is a symbolic model checker for security protocol that combines two concepts based on lazy and demand-driven search. First, lazy data type is a simple way of building efficient on-the-fly model checker for protocols with large or infinite state spaces. The second concept is the integration of symbolic techniques and optimizations to model a lazy Dolev-Yao intruder with the actions generated based on demand-driven concept.

CL-Atse [8] tool is a versatile automatic analyser for the security of cryptographic protocols based on Object-Oriented Categorical Abstract Machine Language (Ocaml). The designed protocol input which is the IF format is specified as a set of rewriting rules and constraint solving techniques are used to model all the states of the users. Then, types of security attacks toward the designed protocol are decided based on Dolev-Yao intruder model capabilities. Algebraic properties of operators like XOR can be modelled using Cl-Atse tool due to a complete modular unification algorithm.

The fourth backend of AVISPA tool is TA4SP which is an approach based on rewriting on regular tree languages to the security protocols verification and is based on unbounded session. Using TA4SP, the intruder knowledge is predicted by using regular tree languages and rewriting to produce under and over approximations. Figure 19 summarize AVISPA concepts with the all four back ends tool.

The proposed protocol for mobile e-health authentication is firstly written in HLPSL coding and then it is converted to an intermediate format. Next, the protocol will be validated by either one of these four backend tools for security analysis. This research considers only two backend tools for analysing the security of mobile e-health authentication protocol which are OFMC and Cl-Atse. It is due to capabilities to completely interpret and validate the protocol coding in a proper way.

Formal analysis on the mobile e-health authentication protocol is based on the developed EHEART application network model presented in Fig. 20 where the e-health sensor node and a mobile device request to be connected with the e-health server. Mobile device acts a gateway for the mobile e-health authentication system which only connects and transfers the data from the e-health sensor node to the e-health server.

In this formal security analysis, three different scenarios are designed for the mobile e-health authentication protocol to be formally analysed in order to get a detailed and thorough security analysis. The following scenarios are:

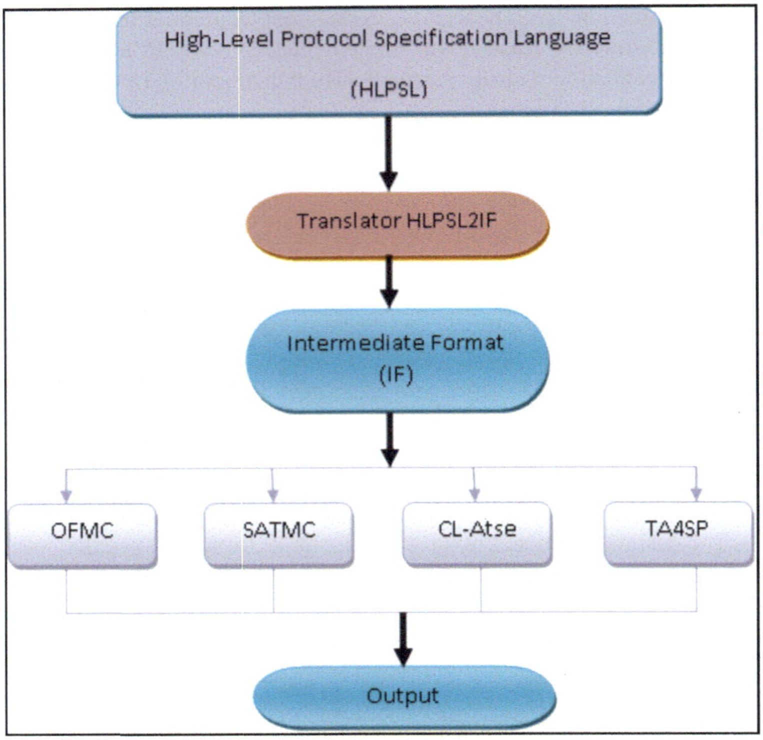

Fig. 19 AVISPA tool structure

Fig. 20 Mobile e-health network model

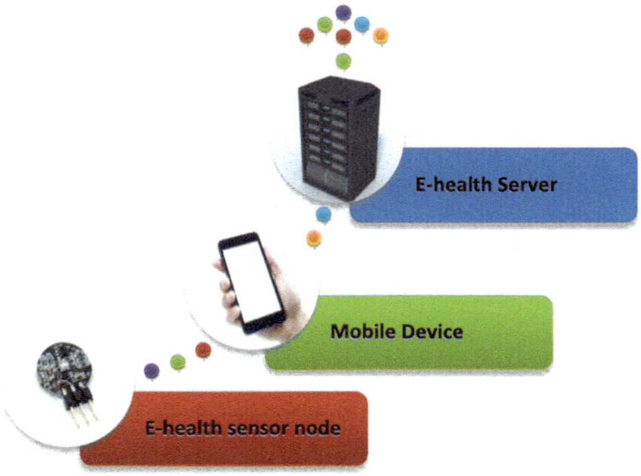

- A valid e-health sensor node attempts to join the e-health network and its session identity is available.
- An illegal e-health sensor node attempts to join the e-health network and its session identity is not available.
- An illegal e-health sensor node attempts to join the e-health network and its session identity is available but having a different unique identity.

Each scenario is validated using the AVISPA backend tool to verify the protocol functionality and its security.

7 Conclusion

This chapter has presented the solution used in the development of a secure mobile e-health authentication protocol based on several security features. These security features are the key factors in understanding the approach to design a security protocol in mobile e-health and achieving security specifications on the mobile e-health network platform. A fully secure mobile e-health system is believed to be unachievable by any means, however, its security protocol should be designed to increase the difficulty level of being attacked by the adversaries hence it will impose extra delay in timing attack. To validate the practicability and feasibility of the proposed protocol, four analysing security mechanisms have been discussed. The following chapter will present a thorough detail on the framework of mobile e-health authentication with the designed mobile e-health authentication protocol.

References

1. Goncalves, F., Macedo, J., Nicolau, M.J., Santos, A.: Security architecture for mobile e-health applications in medication control. In: International Conference on Software, Telecommunications and Computer Networks, SoftCOM 2013, pp. 1–8, Primosten (2013). https://doi.org/10.1109/SoftCOM.2013.6671901
2. Adnan, L.H.: Trusted boot process for wireless sensor node platform. Universiti Teknologi MARA Shah Alam (2012)
3. Rivest, R.L., Shamir, A., Adleman, L.: A method for obtaining digital signatures and public-key cryptosystems. Commun. ACM **21**(2), 120–126 (1978). https://doi.org/10.1145/359340.359342
4. Chatterjee, S., Das, A.K.: An effective ECC-based user access control scheme with attribute-based encryption for WSN. Secur. Commun. Netw. **9**(22), 5968–5974 (2014). https://doi.org/10.1002/sec.1140
5. Salama, D., Kader, H.A., Hadhoud, M.: Studying the effects of most common encryption algorithms. Int. Arab J. E-Technol. **2**(1), 1–10 (2011)
6. Dolev, D.: On the security of public key protocols. IEEE Trans. Inf. Theory **29**(2), 198–208 (1983). https://doi.org/10.1109/SFCS.1981.32
7. Basin, D., Mödersheim, S., Vigano, L.: OFMC: a symbolic model checker for security protocols. Int. J. Inf. Secur. **4**(3), 181–208 (2005)
8. Turuani, M.: The CL-Atse protocol analyser. Rta, pp. 277–286 (2006)

Protocols for Mobile E-Health Authentication

Nazhatul Hafizah Kamarudin, Yusnani Mohd Yussoff, and Tutut Herawan

Abstract This chapter presents the development of a passwordless authentication framework for mobile e-health systems, designed to address security vulnerabilities and enhance data privacy. The framework aims to replace traditional password-based methods, which are often inadequate for mobile health applications due to security limitations and usability concerns, particularly for elderly users. The research introduces the EHEART application, a unique non-regenerated identity generation method for e-health sensor nodes, a two-tier network architecture, and the EHEART authentication protocol. These elements collectively enhance security, minimize reliance on third-party servers, and streamline user access. The EHEART application, leveraging Near Field Communication (NFC), provides a secure and seamless platform for user authentication by activating only when validated by an NFC ring. This approach reduces energy consumption and communication overhead while ensuring that only authenticated users gain access. The framework's two-tier architecture and identity-based encryption eliminate third-party server dependencies, enabling secure data transmission directly from mobile devices to e-health servers. The authentication protocol employs lightweight cryptographic techniques and formal security validation, showcasing resilience against common attacks such as replay and cloning. Results indicate that the proposed framework significantly improves data security and operational efficiency in mobile e-health systems. The use of a unique identity for each sensor node and elimination of passwords enhance security and usability. This passwordless solution contributes to advancing mobile e-health, offering a scalable

N. H. Kamarudin (✉)
Center for Cyber Security, Faculty of Information Science and Technology, Universiti Kebangsaan Malaysia, Bangi, Malaysia
e-mail: nazhatulhafizah@ukm.edu.my

Y. Mohd Yussoff
School of Electrical Engineering, College of Engineering, Universiti Teknologi MARA, Shah Alam, Malaysia

T. Herawan
Department of Information Systems, Faculty of Computer Science and Information Technology, University of Malaya, Kuala Lumpur, Malaysia

N. H. Kamarudin et al. (eds.), *Beyond Passwords: Protecting Healthcare with Authentication Security*, Information Systems Engineering and Management 51,
https://doi.org/10.1007/978-3-031-92247-3_4

and reliable approach that meets the constraints of healthcare environments while improving accessibility and protecting patient data.

Keywords Passwordless authentication · Mobile e-health security · Identity-based encryption · Near field communication (NFC) · Authentication protocol

1 Introduction

The main reason in developing this framework of mobile e-health authentication is to provide a secure and reliable method in validating an e-health sensor node to join the e-health network. Besides implementing a secure authentication method, this framework focuses on removing the e-health network barrier and provides a seamless and efficient communication in mobile e-health by eliminating the e-health service provider role. In addition, the framework is designed in order to reduce the security attacks that occur during the mobile e-health network through cryptographic protocol enhancement. People do realize that it is quite impossible for having absolute security in e-health system however the system administrator such as the healthcare organization should adhere to the security policy and compromise with the security design so that it is difficult for the intruders to get access into the e-health network and it takes longer time to break into the system security. Therefore, designing a mobile e-health security protocol in order to protect the system from such attacks is really important. This chapter presents a succinct description on the designed framework of mobile e-health authentication. It comprehensively describes the mobile e-health network architecture and outlines the details on the mobile e-health authentication protocol.

2 Secure Mobile E-Health Authentication

There are four main components that build the framework of a secure mobile e-health authentication. The components are mobile e-health application namely EHEART development as a secure platform, generation of a unique identity for the e-health sensor node, designing a two-tier mobile e-health network architecture and finally the developing a lightweight mobile e-health authentication protocol namely EHEART authentication protocol. The first component provides a secure application for the mobile e-health platform while the second component prevents the e-health sensor node duplication or cloning attack to protect the e-health sensor node identity. Two-tier mobile e-health network architecture on the other hand develops a secure communication channel by eliminating the third party server in mobile e-health network. Finally, the EHEART protocol acts as an access control scheme to protect the mobile e-health network from an invalid e-health sensor node and thus validate the security of the whole network. The subsequent sections discuss the details on each framework components.

3 EHEART Application Development

EHEART application is firstly developed to create a secure platform for the mobile e-health network. It is a healthcare application supported by electronic processes and communication with the implementation of the NFC ring. Section "Hardware and Software Description" has discussed in detail on the application development methodology. EHEART application is developed in the mobile device and acts as a secure gateway for the e-health sensor node to the e-health server. The EHEART communication allows a user to send identity information without needing to go through multiple steps setting up a connection. Conventional methods such as username and password scheme have reduced the seamless application ability and caused problems to users especially the elderly users. In EHEART application, NFC ring is used to turn on the mobile e-health and activate the Bluetooth module for e-health sensor node communication. The user needs to touch the NFC ring at the back of the mobile device before accessing the EHEART application (Fig. 1).

Bluetooth module is implemented in EHEART application due to its applicable range which is up to 5–10 m range. Thus, Bluetooth module is more secure to be implemented in the mobile e-health application since it can eliminate the adversaries that are not in the communication range. The EHEART application effectively reduces the time consumption by eliminating the username and password scheme and it also reduces the energy usage since the Bluetooth module will be activated if only the NFC ring is validated. NFC ring control application is used to automatically turn on the Bluetooth module and open the EHEART application through a successful NFC identity verification. Simply choose the EHEART application and hold the ring public inlay to the sweet spot at the back of the mobile phone in order to perform NFC ring identity validation process. The EHEART application is only accessible after the NFC ring authentication validation. After a successful verification, BPM reading will show up on the EHEART application if the pulse sensor data is detected. The EHEART application is then available for further communication process in mobile e-health network. It is expected to effectively reduce the time consumption as well

Fig. 1 Overview of EHEART application

as the power consumption of the mobile device since the Bluetooth module and EHEART application will only be turned on when the NFC ring is authenticated. Figure 2 shows the EHEART application process through NFC validation.

This NFC ring authentication technique is implemented in order to secure the EHEART application and to save the mobile device battery life-time as well. After the EHEART application is closed or the mobile e-health communication has ended, the Bluetooth module will turn off the connection with medical device automatically. The EHEART application has to be installed in the NFC supported platform. If the EHEART application is installed in the unsupported NFC technology device, a pop-up alert will show on the mobile device. Figure 3 shows the pop-up alert for the unsupported NFC technology device.

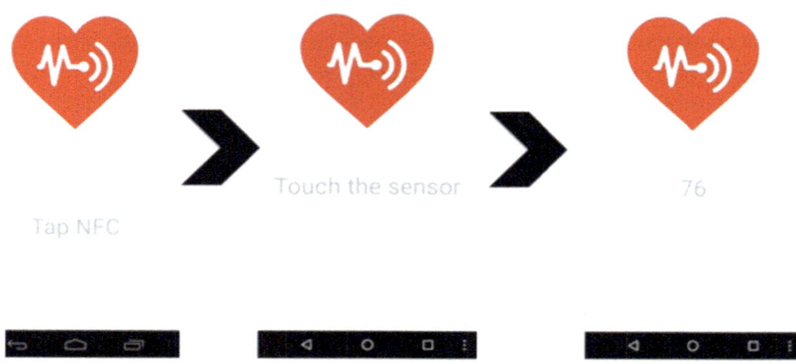

Fig. 2 EHEART application flow process

Fig. 3 Pop-up alert for incompatible mobile device

4 Non-regenerated Unique Identity for the E-Health Sensor Node

Biometric authentication in mobile e-health still comes with several concerns such as higher false error rates in the user access control into the mobile e-health system compared to the unique e-health sensor node identity-based authentication. Plus, biometric authentication can be unlocked by facial recognition software using a picture of a legitimate user and voice recording can surpass most voice authentication [1]. Fingerprints on the other hand are vulnerable to criminals use wherever the user leaves them and it should be handled with an utmost care. Therefore, a non-regenerated unique identity is the right solution to this sensitive issue in the authentication system focusing on the confidential structure like e-health system. In identity-based cryptography protocol, the identity plays a significant role in authenticating and encrypting process where arbitrary string which represents the identity is used as a public key. The identity can be determined from any string such as email address, serial number or project name. However, for mobile e-health application which cannot tolerate duplicate identity, a unique and non-regenerated identity generator is necessary. The unique identity for the e-health sensor node is the outcome from the hash of concatenated unique trust value that is generated during the secure boot process and the NFC ring identity used in the mobile e-health application. To minimize communication overhead, only two bytes from the hash value is extracted as the unique identity of the e-health sensor node platform. Equation 1 depicts the equation used to generate the unique identity of the e-health sensor node (ID_{SN}) in mobile e-health network.

$$ID_{SN} = Hash(uniquetrustvalue||NFCringidentity) \tag{1}$$

The process of generating fresh unique trust identity for the e-health sensor node has been discussed thoroughly in Sect. "Development of a Non-regenerated Unique Identity for E-Health Sensor".

5 Two Tier Mobile E-Health Network Architecture

Therefore, an enhanced e-health network architecture is designed and two-tier network architecture is developed in order to eliminate the function of the third party server or any e-health service provider in the network. It is designed to increase the security level of the e-health system plus to protect the confidentiality of the patient data. Knowledge sharing in e-health system needs to be carefully defined since the existing e-health conducted service is extended outside of its own architecture to the third party public server. Most patients are reluctant to share their medical reports but only to their own doctors and due to that reason, e-health system is not fully operational because of this lack of trust from the people and lack of security enforcement

Fig. 4 Two-tier e-health network architecture

[2–4]. Privacy and confidentiality has to be prioritized in conducting the e-health data transmission. This research proposes a two tier e-health architecture by implementing the designed lightweight IBE protocol namely IBE_EHEART in the mobile device. The IBE_EHEART protocol has eliminated the need for another third party server for key exchange protocol and it uses the e-health server identity as a public key identity for identity-based authentication.

The user or the mobile device has to know the ID of the e-health server since the lightweight IBE algorithm installed in EHEART mobile application will encrypt the data using a public key of the e-health server identity. Thus, the data will be directly transmitted to the e-health server without going through any third party service provider for key generation procedure. Figure 4 shows a direct transmission from the e-health sensor node to the e-health server through the mobile device as a platform gateway.

Lightweight IBE_EHEART algorithm offers a secure data transmission of the authentication message or login information from the e-health sensor node to the e-health server (BS). In Sect. "Identity-Based Encryption Protocol Implementation", it has explained regarding how the minimization of the standard IBE algorithm has resulted to 65% reduction in the BS memory consumption and less complexity in the algorithm implementation where generation of security parameters, private key and master key is done in one single stage instead of two separate phases. The proposed lightweight IBE algorithm has combined the first two phases of the standard IBE protocol while maintaining the outputs. The input of this mobile e-health transmission includes the security parameters, e-health server ID which is the recipient ID, and login information, M. The input message is first encrypted using a public key of BS and the equation for the ciphertext is $C = E_{pub_k}(parameters, ID_{BS}, M)$. Then, the input message is decrypted using a private key of BS to get the login information, M. The decryption equation is $M = d_{k_{BS}}(parameters, ID_{BS}, C)$. Note that login information M ∈ M and the ciphertext C ∈ C.

Lightweight IBE_EHEART algorithm is implemented to eliminate the need for the e-health service provider or any third party server by transmitting the encrypted login information using the public key of the e-health server ID itself. There are three phases of lightweight IBE_EHEART algorithm which is IBE generation phase, encrypt phase and lastly decrypt phase. The IBE generation phase is the combination of setup and extract phase in a standard IBE algorithm implementation.

5.1 IBE Generation Phase

During the IBE generation phase, the e-health server or BS needs to setup the security parameters as well as the master key and extract the private key of the BS.

Step 1: Note that security parameter $k \in Z^*$. Run G on input k in order to generate q, G_1, G_2, and bilinear map $\hat{e} : G_1 \times G_1 = G_2$. Then, a nonce $\alpha \in G_1$ is chosen. A random $s \in Z_q^*$ is selected and set $= \alpha^s$. Cryptographic hash function is chosen $H_1: \{0,1\}^* \rightarrow G_1, H_2: G_2 \rightarrow \{0,1\}n$ for some n. All hash functions are random oracles. The message space for the login information is $M = \{0,1\}n$ and the cipthertext space is $C = G_1 \times \{0,1\}^*$. The output system parameters are $\pi = \{q, G_1, G_2, \hat{e}, n, \alpha, \beta, H_1, H_2, H_3, H_4\}$. The master key is generated by computing $s \in Z_q^*$.
Step 2: To extract the private key of the BS, the IBE algorithm computes $Q_{ID} = H_1(ID) \in G_1^*$. The private key, k is set to be $k_{ID} = (Q_{ID})s$ where s is the master key. Note that a given string $ID \in \{0,1\}^*$.

5.2 Encrypt Phase

The login information m is encrypted before transmitting over the communication channel under the public key ID of the BS. Note that $m \in M$.

Step 1: Compute $Q_{ID} = H_1(ID) \in G_1^*$. A random number $\sigma \in \{0,1\}n$ is chosen and then set $r = H_3\{\sigma, m\}$. Then, the ciphertext is computed as $C = \left(r\alpha, M \oplus H_2\hat{e}(P, P)^{rst}\right)$.

5.3 Decrypt Phase

The ciphertext of the login information is decrypted using the private key of BS which is $k_{ID} \in G_1^*$. Let $c = <U, V, W>$ be a ciphertext encrypted using the BS public key ID. However, if $U \notin G_1^*$, the ciphertext is rejected by the system.

Step 1: Firstly, $V \oplus H_2\left(\hat{e}(k_{ID}, U)\right) = \sigma$ is computed as well as $W \oplus H_4(\sigma) = m$. Set $r = H_3(\sigma, m)$ and then check whether $U = r\alpha$. If $U \neq r\alpha$, the ciphertext is rejected. The output of the login information m is the decryption of c.

Hence, using the lightweight IBE algorithm, the message containing the login information can be transmitted to the BS without the need for a third party server or e-health service provider. Figure 5 shows the details on the IBE_EHEART algorithm development stages.

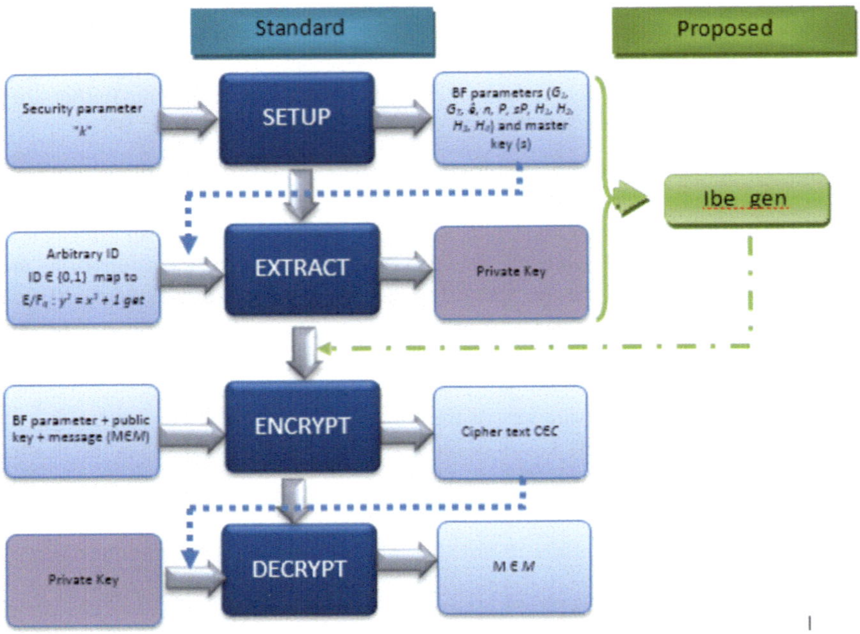

Fig. 5 Lightweight IBE_EHEART algorithm stages

6 Passwordless Authentication Protocol in Mobile E-Health

It is very important to generate a user identity that is unique and non-regenerated for a secure authentication process and identity based authentication has been intensely enhanced to provide such requirements in mobile e-health. Auto identification has been greatly discussed in mobile e-health to minimize the security errors that usually happen during the authentication process. Therefore, by implementing the proposed identity-based authentication in the network system, it contributes in a way to the auto-identification and eliminates the requirement for username and password scheme in most mobile e-health authentication security procedures.

Currently there is a strong need to improve the issues of security and privacy in mobile e-health networking [2]. Therefore, mobile e-health applications should enforce highly strong security authentication protocol in order to protect the system from unauthorized access, wrong identification and mistreatments [5]. The main objective of the EHEART authentication protocol is to provide a passwordless authentication protocol in mobile e-health while ensuring that only valid e-health sensor nodes exist in the network through successful authentication. EHEART

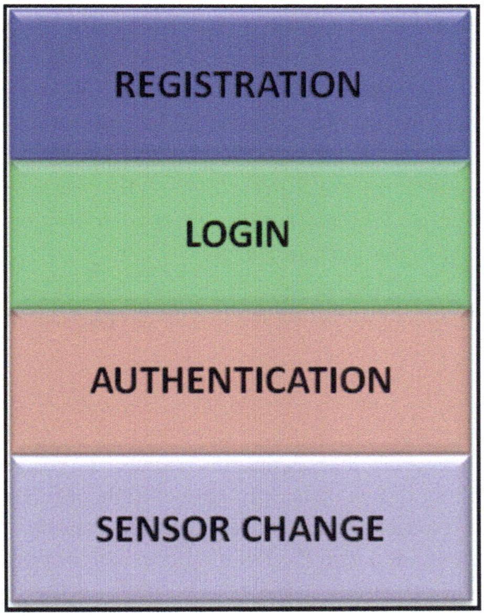

Fig. 6 Stages in EHEART authentication protocol

authentication protocol comprises of four stages starting from security keys generation and security parameters to online e-health sensor node registration and authentication. Figure 6 shows the four stages in EHEART authentication protocol and each stage will be discussed in the following subsections.

6.1 Pre-registration Phase

This phase is used to preload the security parameters keying materials to all e-health sensor nodes and mobile device prior to mobile e-health system deployment. It is executed by the e-health server (BS) in offline mode. It consists of the following steps:

Step 1

The e-health sensor node will go through a secure boot process to generate unique trust value and unique platform identity using Eq. 1. The e-health sensor node unique identity as well as the unique trust value is sent to the e-health server (BS) for further protocol processing. BS will store the values in its e-health database, named as *esnlist*. The unique identity of the e-health sensor node will later become the input to the IBE protocol to produce public and private keys, master key and common security parameters. Before the deployment of the e-health sensor nodes, the e-health server (BS) chooses the following security parameters. The BS generates a finite field q and the parameter p which is a 160-bit prime number, an elliptic curve $E_p(a, b)$ which is the set of all points of $y^2 = x^3 + ax + b\ (mod\ p)$ with the constants $a, b \in Z = \{0, 1, 2, \ldots., p - 1\}$. The BS also generates a unique master key, s for each deployed cluster head where s is only known to the BS. The master key will be used to generate private keys for each e-health sensor node in the mobile e-health network. The public keys associated with the e-health sensor node identity are identified by mapping the identity on the elliptic curve. To obtain a private key, d, the cryptographic hash function of the e-health sensor node unique identity, Q is multiplied with the master key, s presented in Eq. 3.

$$Q_{ID_x} = H_1(ID_X) \tag{2}$$

$$d_x = sQ_{ID_x} \tag{3}$$

The BS also generates a secret key y_i for each of the registered mobile device, where the value is only known to the BS. The BS also selects a unique random value, x_i for each deployed e-health sensor node. All the following security parameters are pre-loaded into the BS memory, *esnlist* before the e-health sensor node deployment. The information list will be used in the authentication process and it is assumed to be valid as the process happen in the control environment which is before the registration phase.

Remark 1 The e-health sensor node needs to be within the e-health server communication range to allow the e-health sensor node to authenticate with the BS and register its unique identity in the *esnlist*.

Remark 2 The e-health server or the e-health base station, BS is assumed to be located in a secure location and it is trustworthy. BS is assumed to be more dominant compared to the e-health sensor node.

Remark 3 The designed mobile e-health application utilizes the Arduino and ARM7 processor with on-SoC memory. It is assumed that the capabilities of the adversaries are not to the extent of being able to extract the valuable information in the protected memory storage of the microprocessor.

Remark 4 The mobile e-health network consists of an e-health sensor node, a mobile device and a BS as the trusted e-health server. This research only considers a two-tier network architecture where e-health sensor node sends the data to the mobile device and then to the BS without the involvement of a cluster head.

6.2 Registration Phase

In the registration phase, a legal user or a valid e-health sensor node needs to register with the BS of the e-health network in order to access real-time communication with the system. This phase is implemented right immediately after the e-health sensor node deployment in mobile e-health system. Upon deployment, the e-health sensor node will boot up and go through the same secure boot process to produce the unique trust value for the registered e-health sensor node. The unique trust value will then be hashed together with the NFC ring identity to create a non-regenerated unique identity of the e-health sensor node. To increase the validity confirmation of the newly registered e-health sensor node identity, the e-health sensor node unique identity will go through the chain-identity attestation (CIA) process to create a session identity to be transmitted over the mobile e-health network. This registration phase has the following steps;

Step 1: User U_i selects random number v_i. After that, user U_i computes $V_i = (ID_{root}|v_i)$ and then sends the registration request message V_i to the mobile device. V_i, ID_{root}, and v_i are stored in the e-health sensor node.
Step 2: Mobile device computes a secret key, y_i and generates a random number w_i. The mobile device then computes $E_i = H\left(V_i \oplus y_i|NFC_{ID_i}\right) \oplus NFC_{ID_i}$
Step 3: BS computes security parameter for user U_i. BS generates the secret master key, s and chooses random value x_i for user U_i. Then, BS computes $W_i = H(w_i \oplus ID_{root})$ using a mobile device random number and $Q_i = W_i \oplus H(x_i) \oplus s$.

In order to become a valid device, a mobile device must register with the BS to obtain authentication IDM and a secret key, y which is shared between the mobile device and the BS.

Remark 1 We assume that the root identity ID_{root} of the e-health sensor node is a very high entropy and strong unique identity so that the e-health sensor node cloning attack will be difficult for an adversary without knowing both trusted value identity and NFC ring identity. According to our threat model, e-health server, BS is considered trustworthy and it keeps all the security parameters as well as the master key of the mobile e-health network. BS will not reveal the master key and the private security parameter keys.

6.3 *Login Phase*

The purpose of the login phase to the mobile e-health network is to validate the user who wants to access the e-health network from a specified mobile device. The following steps are executed:

Step 1: User U_i turns on the e-health sensor node and connected to the EHEART application through the NFC identity verification and Bluetooth module. The e-health sensor node submits its session identity ID_i to the mobile device.
Step 2: The mobile devices then computes $E_i^* = H\left(V_i \oplus y_i|NFC_{ID_i}{}^*\right) \oplus NFC_{ID_i}{}^*$ and checks the condition $E_i^* = E_i$. If this verification does not hold, it means that

U_i has submitted incorrect e-health sensor node identity and the mobile device application will terminate the login phase instantaneously.
Step 3: If $E_i^* = E_i$, the identity is authentic and then mobile device will generate nonce, N to computes $D_i = \left[W_i^* \oplus H(x_i) \oplus H(s) \oplus N\right]$ and $R_i = E_{pub_k}\left[D_i || NFC_{ID_i}\right]$ where R_i is encrypted with a BS public key (pub_k) using IBE lightweight encryption protocol.
Step 4: Finally, mobile device sends login request message $\langle R_i \rangle$ to the BS via a public channel.

6.4 Authentication Phase

A mutual authentication between the mobile device and the BS must be executed. When the login message $\langle R_i \rangle$ is received from the mobile device, the BS performs these following steps:

Step 1: Upon receiving the login message, the BS decrypts the encrypted login message $\langle R_i \rangle$ using its private key, p_k to obtain the security parameters.
Step 2: Then, BS checks $NFC_{ID_i}^* = NFC_{ID_i}$. If there is a match, the mobile device is considered as a legitimate user and BS will accept the connection if the NFC_{ID_i} is valid. Otherwise, this phase is terminated immediately.
Step 3: BS then computes $Q_i^* = D_i \oplus N \oplus H(s) \oplus s$ using its own master key and checks if the received Q_i^* is equal to the retrieved Q_i. If this condition holds true, a secure session key has established and the e-health sensor node has been successfully authenticated.

6.5 Sensor Change Phase

Step 1: If the user wants to change the e-health sensor node in the mobile e-health, the new e-health sensor node needs to re-register in order to generate a new unique identity and get the trust unique value through the secure boot process. During the e-health sensor node change phase, the user is asked to resubmit new e-health sensor node identity in the registration phase. The user U_i will resubmit $V_{i_{new}} = \left(ID_{root_{new}} | v_i\right)$.
Step 2: The E_i value in the mobile device will be updated to $E_{i_{new}} = H\left(V_{i_{new}} \oplus y_i | NFC_{ID_i}\right) \oplus NFC_{ID_i}$.
Step 3: E-health server then computes $W_{i_{new}} = H\left(w_i \oplus ID_{root_{new}}\right)$. Then, the value of W_i is replaced with $W_{i_{new}}$ in the e-health server using the same mobile device. The new e-health sensor node is finally successfully changed in the system.

Remark 1 When the BS needs to revoke a user access rights to the e-health data system, it simply removes the corresponding security parameter pieces of the user in each of the proxy database. Note that the data does not need to be re-encrypted and none of the other user's will be affected since only the security parameters corresponding to the user are removed from the BS database. Since the data does not need to be re-encrypted nor does need to be any key re-distribution, the protocol model is efficient and feasible to be implemented in the mobile e-health system.

7 Identity Based Authentication in Mobile E-Health

Figure 7 depicts the simplified attestation model for newly joined e-health sensor node in the proposed mobile e-health network application. The flow of the processes are numbered chronologically starting from the generation of the e-health sensor node trust value and its unique identity. The EHEART application can be only turned on if the condition of the NFC identity is true. Upon receiving login request from the mobile device, BS will decrypt the login information packet using its private key and compare the e-health sensor node trust value as well as its unique identity. Upon successful authentication, the mobile device will receive acknowledgement packet (ACK) and the e-health sensor node is successfully authenticated.

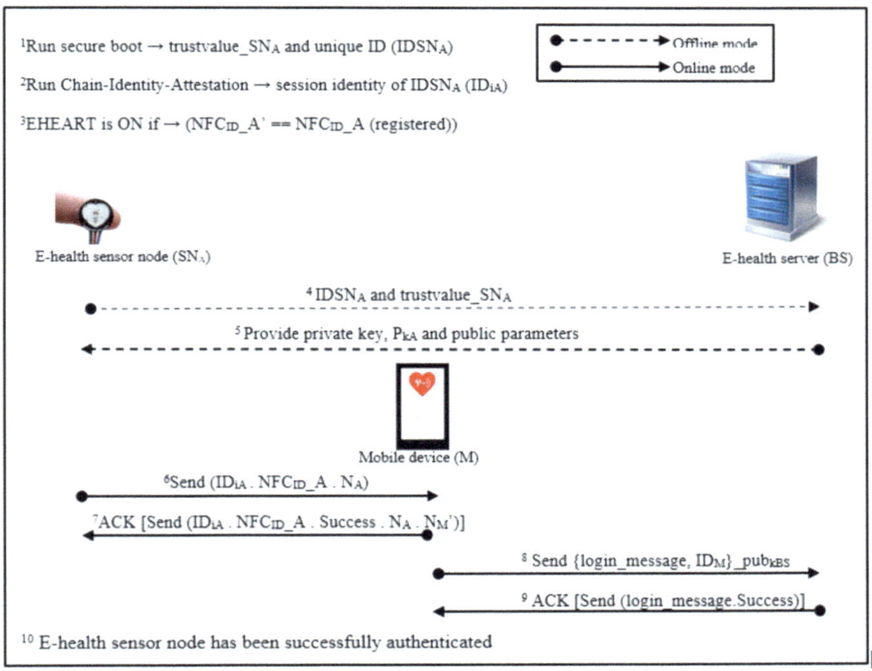

Fig. 7 Simplified attestation model for newly joined e-health sensor node

The authenticated e-health sensor nodes in the mobile e-health network remain in a trusted condition as long as it remains in the ON state. Once the system is rebooted or shut down for any reason, the e-health sensor node needs to re-authenticate with the e-health server or BS again. Failure to get authenticated will lead to e-health sensor node termination process. Upon successful identity based authentication with the BS, the e-health sensor node will successfully join the mobile e-health network and establish secure communication with the BS.

This proposed passwordless identity-based authentication has also considered several efficiency assumptions to avoid resource depletion in the resource constraint for the e-health sensor node. The efficiency assumptions for mobile e-health application are communication overhead, memory overhead, and energy consumption rate.

7.1 Communication Overhead

Communication overhead is referring to the proportion of additional communication payload of data needed outside of data exchange. There are two processes that have been discovered of potentially contributing extra communications in the mobile e-health network. The following briefly explains how this research has minimized the communication overhead in the network.

- *E-health sensor nodes interaction*: E-health sensor nodes only need to register with the BS once during the registration phase assuming no reboots or shutdown. Plus, the e-health sensor node only receives its trust value and unique identity from the BS when it is a newly joining e-health sensor node.
- *Authentication message*: Login information message during the communication transmission channel has been minimized to only one packet message respectively to avoid communication overhead.

Communication data in terms of number of messages and the size of each message in bytes has been minimized in order to reduce the total of communication overhead in the EHEART system.

7.2 Memory Overhead

Memory overhead can incur in two types of overhead which is firstly, the extra time spent to access memory within the BS and secondly, the additional space needed by the BS for its own coding and data structures beyond the memory allocated. The e-health sensor node platform has a considerable amount of memory space and most of the protocol algorithm is implemented on the BS memory space system. The lightweight proposed authentication protocol only requires the memory space that is available to store the common security parameters, private key, and list of trusted e-health sensor node identity. To ensure the feasibility and security of the designed authentication protocol in the resource constraint environment, the issue regarding the location of the list of trusted e-health sensor node identity ($IDSN_{list}$) has to be put into a detailed consideration. It is then decided to let the mobile device check the authenticity of the requester e-health sensor node with the BS since a list of all trusted e-health sensor nodes identity will be installed in the BS memory. It might lead to an extra communication overhead. However, if the list of trusted e-health sensor node identity is installed in the mobile device as well, it will contribute to an additional memory requirement in the mobile device which will reduce its performance and an extra security element needs to be implemented to protect the mobile device.

Moreover, the size of the list of trusted e-health sensor node identity kept in the BS memory space is less than 2 bytes per e-health sensor node which is not a burden for the BS memory space. For instance, applications like EHEART might only have maximum 10 e-health sensor nodes per network and this the highest memory size will be only around 20 bytes. Equation 4 represents the total bytes of memory size of trusted $IDSN_{list}$ where N is the number of the e-health sensor nodes in the EHEART network.

$$Memory\ size\ of\ trust\ IDSN_{list}(bytes) = 2N - 2 \tag{4}$$

7.3 *Energy Consumption*

Communication process has consumed the highest energy consumption compared to the other processes because of a high transmission and reception power. In order to reduce significant energy consumption for the designed protocol, the number of communications between the e-health sensor node to mobile device and mobile device to BS authentication has to be minimized. Plus, another aspect that has been put into consideration is the extra computational process in the cryptographic protocol implementation. The designed framework for mobile e-health authentication has been intended to transfer the heavy process such as security parameters and key generation to BS during the pre-registration phase. Although it creates an additional process before the registration phase, this pre-registration phase can reduce the power consumption of the e-health sensor node during the authentication process.

8 Conclusion

The findings demonstrate the framework's effectiveness in providing secure data transmission, resilience against cloning and replay attacks, and improved energy efficiency. The use of identity-based encryption and lightweight cryptographic techniques further reduces the complexity and memory requirements, making it suitable for resource-constrained environments. The proposed solution not only enhances security but also promotes user accessibility, aligning with the increasing demand for reliable and scalable mobile e-health systems.

References

1. Hadid, A.: Face biometrics under spoofing attacks: vulnerabilities, countermeasures, open issues, and research directions. In: IEEE Computer Society Conference on Computer Vision and Pattern Recognition Workshops, no. Cmv, pp. 113–118 (2014). https://doi.org/10.1109/CVPRW.2014.22
2. Thilakanathan, D., Chen, S., Nepal, S., Calvo, R., Alem, L.: A platform for secure monitoring and sharing of generic health data in the cloud. Elsevier Future Gener. Comput. Syst. **35**, 102–113 (2014). https://doi.org/10.1016/j.future.2013.09.011
3. Guo, L., Zhang, C., Sun, J., Fang, Y.: A privacy-preserving attribute-based authentication system for mobile health networks. IEEE Trans. Mob. Comput. **13**(9), 1927–1941 (2014). https://doi.org/10.1109/TMC.2013.84
4. Kamarudin, N.H., Yussoff, Y.M., Hashim, H.: Two-tier e-health monitoring system. In: WSEAS Applied Computational Science (ACACOS), pp. 154–158, Kuala Lumpur (2015)
5. Goncalves, F., Macedo, J., Nicolau, M.J., Santos, A.: Security architecture for mobile e-health applications in medication control. In: International Conference on Software, Telecommunications and Computer Networks, SoftCOM 2013, pp. 1–8, Primosten (2013). https://doi.org/10.1109/SoftCOM.2013.6671901

Performance Analysis of Mobile E-Health Authentication System

Nazhatul Hafizah Kamarudin, Yusnani Mohd Yussoff, and Tutut Herawan

Abstract This chapter presents a comprehensive performance analysis of a passwordless authentication framework, specifically designed for mobile e-health systems. The primary goal is to assess the feasibility and efficiency of the EHEART authentication protocol by examining key metrics, such as energy consumption, memory utilization, and computational overhead. Recognizing the limitations of password-based methods in healthcare, especially regarding security and usability, the study focuses on developing an optimized, identity-based encryption protocol that meets the constraints of mobile health applications. The analysis includes evaluating the energy efficiency of critical processes like secure boot, encryption, and authentication on an Atmega328p microprocessor. Results indicate that the EHEART protocol achieves substantial reductions in energy consumption, particularly in secure boot and data transmission phases, making it well-suited for low-power devices. A comparison with other protocols demonstrates that EHEART maintains competitive energy efficiency while eliminating the need for external cryptographic hardware, thereby reducing costs and complexity. In terms of memory utilization, the lightweight identity-based encryption (IBE) algorithm used in EHEART reduces memory usage by over 65% compared to conventional IBE implementations. This reduction enables deployment on devices with limited memory resources, enhancing scalability and practicality for large-scale e-health networks. Additionally, the framework's architecture eliminates third-party servers, directly connecting sensor nodes to mobile devices and e-health servers, which minimizes communication overhead and enhances data security. The findings confirm that the EHEART framework offers

N. H. Kamarudin (✉)
Center for Cyber Security, Faculty of Information Science and Technology, Universiti Kebangsaan Malaysia, Bangi, Malaysia
e-mail: nazhatulhafizah@ukm.edu.my

Y. Mohd Yussoff
School of Electrical Engineering, College of Engineering, Universiti Teknologi MARA, Shah Alam, Malaysia

T. Herawan
Department of Information Systems, Faculty of Computer Science and Information Technology, University of Malaya, Kuala Lumpur, Malaysia

N. H. Kamarudin et al. (eds.), *Beyond Passwords: Protecting Healthcare with Authentication Security*, Information Systems Engineering and Management 51,
https://doi.org/10.1007/978-3-031-92247-3_5

a viable solution for secure, efficient, and user-friendly authentication in mobile e-health systems. By addressing critical concerns of energy consumption, memory constraints, and user accessibility, this framework supports the growing adoption of digital health technologies while safeguarding patient data and improving the user experience.

Keywords m-health authentication · Passwordless authentication · Energy-efficient encryption · Identity-based encryption (IBE) · Memory optimization

1 Introduction

This chapter comprehensively presents and discusses the performance analysis obtained from the analysis of the proposed framework for the mobile e-health authentication protocol. The main objective of this chapter is to prove the feasibility of the designed framework and the authentication protocol in the EHEART mobile application. The performance of the proposed framework is assessed thoroughly on the protocol feasibility and the efficiency.

2 Protocol Feasibility

Protocol feasibility or protocol practicality is evaluated in terms of energy consumption and memory utilization. This section will present a detail analysis on the energy consumption and memory utilization of the proposed framework of the EHEART authentication protocol in mobile e-health network.

2.1 Energy Analysis

In this subsection, the energy analysis of selected background processes in the framework development is presented. The energy consumption of the designed passwordless authentication for EHEART application is also presented for further discussion.

The following results were obtained from the energy analysis performed on Atmega328p development board and the reason for implementing the Atmega328p microprocessor in the e-health sensor node platform is due to its higher security features that can fulfil the mobile e-health specifications and it has been designed for low power applications. The Atmega328p microprocessor runs at 6 mA, 3.3 V with an external frequency of 16 MHz. As encryption and communication process as well as the authentication process utilize most of the energy consumption in the designed framework, thus this study only considers the amount of energy used in those processes. Since e-health sensor nodes are resource-constrained, the energy cost analysis during the login and authentication phase also needs to be put into consideration.

As part of benchmarking, this research compares the result with [1–4] that utilizes identity-based authentication and use public key encryption in mobile e-health authentication scheme. Referring to Table 2, the development microprocessor board took around 0.5 s to complete the secure boot process which includes evaluating the unique trust value of the e-health sensor node. Energy consumption during the secure boot process is 36 mJ and it is so much lower compared to the other trusted boot process. Significant reduction in total energy consumption for the proposed secure boot process in EHEART authentication protocol is due to the minimized command execution.

However, the total energy per bit for the proposed encryption process in EHEART protocol is around 22.5 μJ/bit which is four times higher than the hardware based encryption process because it is fully implemented in software based encryption. Even though the total energy utilized in the proposed encryption process is considerably a little bit higher compared to [5, 6], but it does not burden the e-health sensor node with any external crypto microprocessor chip on the e-health sensor node platform.

Another widely used process in this EHEART authentication protocol is the secure hash algorithm (SHA). During the secure boot process, the algorithm will capture one of the component identities of the e-health sensor node and use it to generate the unique identity. SHA algorithm then is used to generate the 20 bytes platform unique identity. However, in order to minimize the communication overhead, the size of the unique identity has been limited to only 2 bytes.

Table 1 shows the benchmark analysis of the SHA algorithms. SHA-256 algorithm is proved to be the lowest time delay consumption.

Referring to Table 2, the comparison on the energy consumption between the EHEART and other works presented [5–7] that utilizes Atmel microcontroller indicate that SHA-256 is feasible in the implementation of the e-health sensor node with 32-bit processor. The energy per byte for SHA-256 is 0.028 μJ which is very low and thus can minimize the energy cost in the security key generations. The authentication protocol is implemented in order to run the authentication process securely. Tate pairing is an internal algorithm that is used in the generation of public and private key in IBE cryptography. Authentication process as well as tate pairing

Table 1 Benchmark analysis of the SHA algorithms

SHA	Time (μs/byte)	Compression bit	ROM size (K)	Ciphertext size	Round
SHA-160	0.41	32	1.29	160	80
SHA-256	0.39	32	1.59	256	64
SHA-512/384	0.83	64	3.3	512/384	80

Table 2 Comparison of energy consumption on selected major processes

	Energy consumption							
	Delay (s)				Energy (J)			
Process	EHEART (proposed)	Wen Hu et al. [6]	Najwa et al. [5]	Anthony et al. [7]	EHEART (proposed)	Wen Hu et al. [6]	Najwa et al. [5]	Anthony et al. [7]
Secure boot	0.5	0.5	2.442	0.5	0.0036	0.0054	5.746	0.008
Encryption	0.05	0.055	3.579	2.45	22.5×10^{-6}	5.4×10^{-6}	8.034	0.0398
SHA2	0.39×10^{-6} per byte	18×10^{-6}	NA	0.566	0.028×10^{-6}	5.4×10^{-6}	NA	0.0092
Authentication protocol	4.05	NA	4.859	NA	0.292	NA	11.82	NA

consumed the highest energy consumption as shown in Table 2 due to their complicated computation. However, this proposed work obtained 0.292 J which is lower than the result obtained by Najwa et al. [5, 8] (not included in the table) that utilized ARM7 processor. Thus, it validates the reason for using a low power processor for mobile e-health application and the elimination of the tate pairing algorithm in the online phase of the EHEART authentication process. Note that, the proposed authentication protocol for mobile e-health does not require any packet exchange and due to application of ECC and Boneh-Franklin algorithm, the energy cost of data transmission has been greatly reduced.

The authentication protocol utilized the most energy consumption followed by the secure boot and encryption process. The SHA 256 energy consumption is negligible and can be ignored.

Figure 1 shows the percentage of energy analysis distribution for selected major processes running in the microprocessor. Energy utilization analysis is calculated and presented in order to realize the complex computational process of a secure authentication protocol.

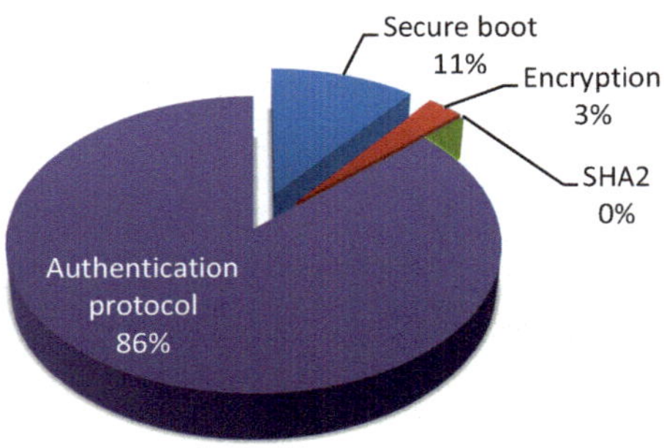

Fig. 1 Energy analysis in EHEART for selected major process

Power consumption analysis of the embedded system is calculated as well to discuss the power consumption on EHEART application and the mobile phone battery lifetime. Calculation on power consumption per hour is done as well as voltage analysis. 9 V is the initial voltage value before the data transmission and the voltage drop during the data transmission is about 3 V. The analysis of the voltage is done to produce a stable voltage using the voltage regulator LM7805 and CJP1117. Table 3 depicts the results of the input voltage as well as output voltage on step-down voltage regulators.

Voltage regulator LM7805 is used to supply 5V to the EHEART microcontroller and voltage regulator CJP1117 is used to supply 3.33 V to the Bluetooth module. It is shown that in graph Fig. 2 that the both voltage regulators have successfully produced a stable voltage for each component in EHEART application.

Then, power consumption analysis is done given that voltage is 9 V and current (I) is 4.54 mA based on the simulation result.

$$P = IV$$

$$P = 4.54\,\text{mA} \times 9V = 40.86\,\text{mWatt} \tag{1}$$

$$P/\text{hour} = 40.86\ \text{mWatt} \times 3600\ \text{seconds} = 147.096\ \text{mAh} \tag{2}$$

Then, the result for power consumption per hour is used to do the battery lifetime analysis for the mobile device. Based on the results in Table 4, it can be assumed that using a higher mobile device battery capacity, EHEART application will last much longer since the power consumption of the EHEART embedded devices is only 40.86 mW. Plus, by only using 400 mAh mobile device battery capacity, the EHEART

Table 3 Results of step-down module for input and output voltage

Battery (V)	LM7805 voltage regulator		CJP1117 voltage regulator	
	V_{in}	V_{out}	V_{in}	V_{out}
9.0	9.0	5.06	5.06	3.3
8.9	8.9	5.06	5.06	3.3
8.8	8.8	5.06	5.06	3.3
8.7	8.7	5.06	5.06	3.3
8.6	8.6	5.06	5.06	3.3
8.5	8.5	5.06	5.06	3.3
8.4	8.4	5.06	5.06	3.3
8.3	8.3	5.06	5.06	3.3
8.2	8.2	5.06	5.06	3.3
8.1	8.1	5.06	5.06	3.3
8.0	8.0	5.06	5.06	3.3

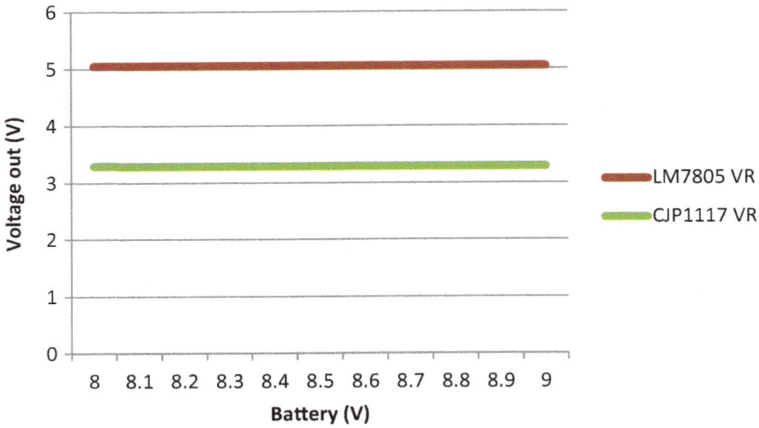

Fig. 2 Graph on output voltage for LM7805 and CJP1117 voltage regulators

application can be used up to 2 h and 43 min non-stop since its power consumption per hour is 147.096 mAh. It can be referred to the Eq. 2. Table 4 shows the theoretical analysis on the time availability of the EHEART embedded system during the ON active condition. Using the normal 3000 and 4000 mAh battery capacities in mostly mobile device, the EHEART application can be used up to 27 h non-stop in ON condition depending on the mobile device battery capacity. Details calculation on the EHEART time availability is presented below where the mobile device battery capacity is 400, 3000, and 4000 mAh.

$$EHEART\ time\ availability\_1 = \frac{400\,\text{mAh}}{147.96\,\text{mAh}} = 2.7193\,\text{hours} \tag{3}$$

$$EHEART\ time\ availability\_2 = \frac{3000\,\text{mAh}}{147.96\,\text{mAh}} = 20.27\,\text{hours} \tag{4}$$

$$EHEART\ time\ availability\_3 = \frac{4000\,\text{mAh}}{147.96\,\text{mAh}} = 27.03\,\text{hours} \tag{5}$$

Compared to HEALTH-Dr application [9] and MobiHealth application [10, 11] which can last no longer than 8 h during the ON active condition using the normal mobile device battery capacity. Several other mobile e-health applications such as CustoMed application [12] can last up to only 9 h during the ON condition without any security implementation in the mobile e-health application.

Table 4 EHEART application analysis on the time availability

Mobile device battery capacity (mAh)	Available period in ON state (h)
400	2.7193
1000	6.758
3000	20.27
4000	27.03

2.2 Memory Utilization Analysis

In this subsection, the memory utilization analysis in security parameters storage of the proposed mobile e-health authentication protocol is evaluated. The key pre-registration scheme proposed by Awasthi et al. [13], Du et al. [14, 15], and Szczechowiak et al. [16] are used for memory analysis comparison.

In the proposed EHEART authentication scheme, each sensor and mobile device is preloaded with its private key, ID and public key. The e-health server also needs to store a secure ID_{list} for a successful session key establishment between the e-health sensor nodes and the e-health server. Total number of a preloaded keys and parameters that contribute to memory utilization and storage overhead in EHEART authentication protocol is defined as the Eq. 4. Table 5 depicts the terminology of the memory utilization analysis on selected keys distribution scheme.

$$Total\ no.\ of\ preloaded\ keys = 2 + (2N - 2) \tag{6}$$

Szczechowiak et al. scheme [16] produces a constant number of keys because there is no private key is stored in the e-health sensor node. Szczechowiak et al. scheme in the first phase preloaded all the necessary parameters to the e-health sensor nodes during the offline mode before the network implementation. Du et al. Asymmetric Pre-Distribution scheme [14] and Routing Driven Key-Management scheme [15] as well as Awasthi et al. scheme [13] on the other hand result in a very significant increase in the number of keys as the network size increases.

The need for a proper key distribution scheme in the e-health system implementation is essential to avoid the linear increment of total number of keys in the e-health network in order to allow network scalability in the mobile e-health system.

Table 5 Terminology of the memory utilization analysis

M	Number of high end e-health sensor nodes which is usually the cluster head or a mobile device
N	Number of low end e-health sensor node
m	Number of preloaded keys in e-health sensor nodes
x, y	Shared key between e-health sensor nodes and mobile device

Table 6 shows the memory utilization space overhead analysis for each keys distribution scheme.

Memory utilization comparison between the described schemes is done and the parameters are set as $M = 30$, $m = 100$, $x = 500$, and $y = 20$. The outcome from the graph analysis in Fig. 3 shows the feasibility of the proposed authentication scheme in mobile e-health system. From the analysis, it can be seen that EHEART protocol requires less storage space for preloaded keys compared to other protocols.

The memory utilization space overhead increases with the increasing number of pre-loaded keys. Szczechowiak et al. scheme is independent of the network size and thus introduces significant savings in memory utilization space overhead since it does

Table 6 Memory utilization space overhead analysis

Keys distribution scheme	Memory utilization
EHEART	$2 + (2N - 2)$
Awasthi et al. [13]	$m(M + N)$
Du et al. (Asymmetric Pre-Distribution) [14]	$xM + yN$
Du et al. (Routing Driven Key-Management) [15]	$M(N + 3) + 2N$
Szczechowiak and Collier scheme [16]	$3M$

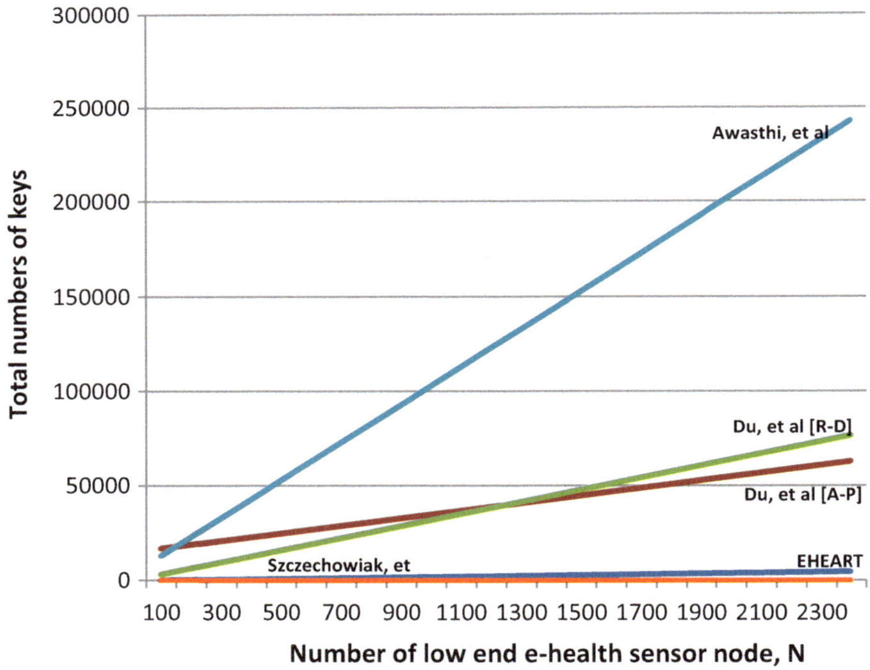

Fig. 3 Comparison on the total number of preloaded key storage space

Table 7 Memory size comparison between current IBE and proposed lightweight IBE_EHEART process

	Current IBE algorithm	Proposed lightweight IBE algorithm
Process	Memory size (MB)	
Encrypt	35.9	12.8
Decrypt	35.2	11.9
Total	71.1	24.7

not store any private key in the e-health sensor node. EHEART on the other hand has only a slight increasing number of memory utilization space overhead since in each e-health sensor node is preloaded with two keys only which are its public key ID together with its private key.

Further memory size analysis comparison is done between the current IBE algorithm [17] and the proposed lightweight IBE algorithms in EHEART authentication process during encrypt and decrypt phase. To realize the computational complexity of IBE algorithm, Table 7 tabulates the total usage of memory size for the IBE algorithms.

The result obtained from the minimization of the memory size of IBE algorithm shows that it is suitable to be implemented on EHEART application since there is 65.2% of memory size reduction compared to the existing IBE algorithm. The existing IBE algorithm has used up to 71.1 MB memory size while the proposed lightweight IBE algorithm only use about 24.7 MB for its total memory size usage. This research has successfully produced a practical lightweight IBE algorithm thus making it feasible for EHEART application as well as Atmega328p microprocessor specifications and constraints.

The Fig. 4 presents the results obtained from the memory size comparison between the current IBE algorithm and the lightweight IBE algorithm in EHEART framework. It shows that lightweight IBE algorithms have successfully reduced more than 65% of the memory size in the existing IBE algorithms hence making it more practicable for mobile e-health that requires limited memory usage in its application. Upon successful development for lightweight identity-based encryption in mobile e-health system, the memory size of IBE encryption and decryption has been greatly minimized in order to facilitate the constraint in Atmega328 microprocessor.

3 Protocol Efficiency

Further analysis on the IBE_EHEART is done to evaluate the protocol efficiency in mobile e-health system. To erase the drawbacks of password based authentication schemes, several proposed authentication schemes have suggested to use public key cryptosystem [3]. A public key infrastructure (PKI) is a set of components that manages keys and certificates. PKI uses two keys merely public and private which

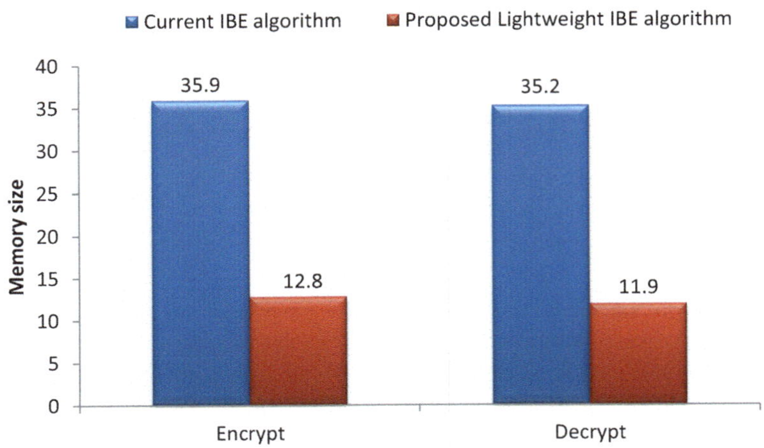

Fig. 4 Comparison on existing IBE algorithm and lightweight IBE algorithm

the public key is available to the public in order to decrypt the data and only the one possessing the private key can decrypt the data. PKI based authentication has long been used in communication protocols to eliminate man-in-the-middle attack. However, most of the authentication protocols do not facilitate the limited constraints of mobile e-health application. Thus, this proposed identity-based encryption process has been enhanced and developed to be lightweight for mobile e-health application system. In order to facilitate the constraint in sensor network application from e-health sensor node to the mobile device, the two IBE stages are combined into only one stage which is IBE_generate and it has significantly reduced the memory usage in the e-health system for more than 50%. With the high memory reduction in the lightweight IBE_EHEART implementation, it is now possible to implement the algorithm in mobile e-health application. Table 8 shows the comparison between the current IBE algorithms which have two IBE stages and proposed IBE_generate in the lightweight IBE_EHEART algorithm which only has one stage combined for security parameters generation.

It is further shown in Fig. 5 regarding the memory size consumption analysis between the current IBE which includes IBE_setup and IBE_extract with with IBE_ generate phase in IBE_EHEART.

Table 8 Comparison between the identity-based encryption source code

Process	File size (bin) KB	Library size (KB)	RAM (KB)
IBE_setup	165	139.4	12.8
IBE_extract	167	140.0	12.8
Total	322	279.5	25.6
IBE_generate (proposed)	166	138.4	12.8

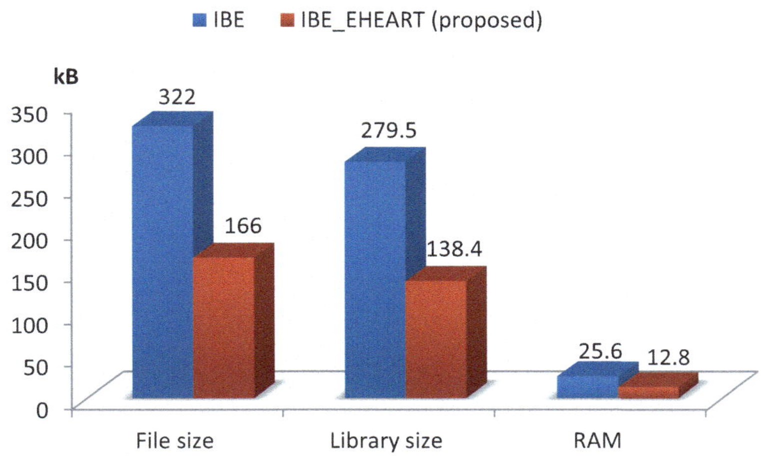

Fig. 5 Memory size consumption for current IBE and IBE_EHEART

Next, the functionality features and system performance of our proposed EHEART authentication protocol scheme is compared with those other related mobile e-health authentication schemes. For the performance comparison and standardization purposes, the notations listed in Table 9 are used. The efficiency of the proposed protocol in terms of communication cost and computation load are systematically evaluated based on Table 10.

For assymetric cryptosystem, the computational time for executing encryption and decryption is taken as that for a modular exponentiation operation. For simplicity, the time taken for one MAC operation is considered for one hashing operation. The performance of our proposed scheme is compared with other related mobile e-health authentication schemes for login and authentication phase. It is also assumed that the time for executing an XOR operation is negligible. Note that some of the results are derived from [4, 18].

Figure 6 presents the rough estimation of computational cost among the existing authentication schemes in mobile e-health system. Based on the comparison made

Table 9 Notations used for the computational complexity

Symbol	Description
T_P	Time for performing pairing
T_M	Time for performing elliptic curve point multiplication
T_I	Time for performing modular inverse
T_h	Time for performing one-way hashing operation
T_x	Time for performing an XOR operation
T_E	Time for performing encryption
T_D	Time for performing decryption

Table 10 Computational cost comparison of our scheme with existing schemes

Schemes	Login phase	Authentication phase
Zhou et al. scheme [19]	$3T_M + T_I + T_h + 1T_E + 1T_D$	$3T_M + T_I + T_h + 1T_E + 1T_D$
Tan et al. scheme [20]	$4T_h + 2T_x + 1T_E$	$3T_h + 1T_x + 1T_D$
Zuowen Tan scheme [21]	$4T_h + 3T_x + 2T_M$	$4T_h + 1T_x + 3T_M$
Mishra et al. scheme [3]	$3T_h + 2T_E$	$6T_h + 4T_E$
Chatterjee et al. scheme [22]	$7T_h + 4T_M + 3T_x + 1T_E$	$5T_h + T_M + 1T_E + 1T_D$
Das scheme [23]	$2T_h + 3T_x$	$5T_h + 2T_x$
Lee and Hsu scheme [24]	$5T_h + 4T_x$	$5T_h + T_x$
Awasthi and Srivastava scheme [13]	$2T_x + 3T_h + 4T_x + 1T_E + 1T_D$	$4T_h + 4T_x$
Arshad and Nikooghadam scheme [25]	$3T_h + 5T_x + T_M$	$8T_h + 6T_x + 2T_M$
Ashok scheme [4]	$5T_h + 4T_x + 2T_M$	$2T_h + 2T_x + 3T_M$
EHEART scheme	$4T_h + T_M + T_P + 1T_E$	$3T_h + T_M + T_P + 1T_D$

with the other existing schemes, it can be referred that EHEART authentication scheme is acceptable and within the computational cost range among the other mobile e-health authentication protocols. Even though the EHEART authentication scheme introduces the unique and trust element into the network protocol, the computational cost is found to be comparable with other existing schemes that do not implement the unique identity and trust validation.

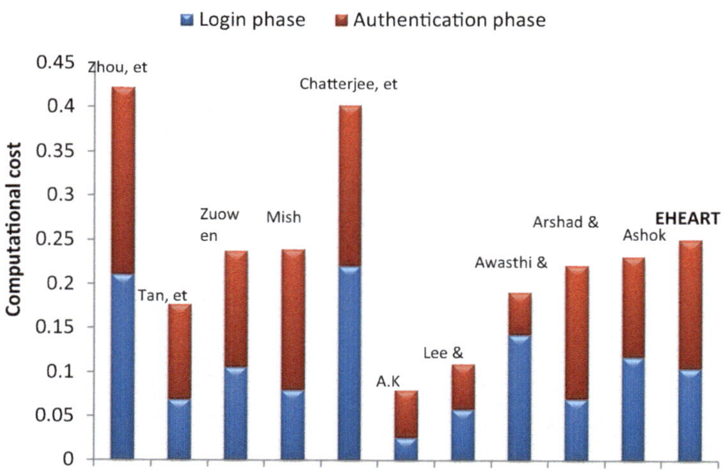

Fig. 6 Computational cost comparison

4 Conclusion

Threats in the current mobile e-health networking appear in a variety of forms and adversaries which are embedded with different intentions. There are numerous types of attacks that aim to steal or modify the valuable asset or login information in the mobile e-health system. Successful attacks depend on the total vulnerabilities and also the weakness level of the vulnerabilities surrounding the valuable entity, which is referring to the e-health sensor node. This proposed EHEART authentication protocol is developed to provide a secure authentication framework in the network. Memory analysis, protocol feasibility and efficiency as well as the security analysis are done to analyse the proposed framework. Thus, based on the results analysis, the proposed authentication protocol is proven to be secured from replay attack and sensor node cloning attack. In addition, the proposed authentication protocol has successfully reduced up to 65% reduction in memory size consumption and thus it is efficient to be implemented in the mobile e-health system.

References

1. Maitra, T., Giri, D.: An efficient biometric and password-based remote user authentication using smart card for telecare medical information systems in multi-server environment. J. Med. Syst. **38**(12) (2014). https://doi.org/10.1007/s10916-014-0142-x
2. Odelu, V., Das, A.K., Goswami, A.: An efficient key-management scheme for hierarchical access control in E-medicine system. J. Med. Syst. **36**(4), 2325–2337 (2013). https://doi.org/10.1007/s10916-011-9700-7
3. Mishra, R., Barnwal, A.K.: A privacy preserving secure and efficient authentication scheme for telecare medical information systems. J. Med. Syst. **39**(5), 1–10 (2015). https://doi.org/10.1007/s10916-015-0215-5
4. Das, A.K.: A secure user anonymity-preserving three-factor remote user authentication scheme for the telecare medicine information systems. J. Med. Syst. **39**(3) (2015). https://doi.org/10.1007/s10916-015-0218-2
5. Aaraj, N., Raghunathan, A., Ravi, S., Jha, N.K.: Energy and execution time analysis of a software-based trusted platform module. In: Proceedings-Design, Automation and Test in Europe, pp. 1128–1133 (2007). https://doi.org/10.1109/DATE.2007.364446
6. Shih, W.C., Hu, W., Corke, P., Overs, L.: sefleck: a public key technology platform for wireless sensor networks. In: Proceedings of the 6th ACM Conference On Embedded Network Sensor Systems—SenSys'08, p. 447 (2008). https://doi.org/10.1145/1460412.1460496
7. Marcus, A., Cardei, I., Furht, B.: A mobile device prototype application for the detection and prediction of node faults in wireless sensor networks. J. Netw. Internet Archit. (2014)
8. Doyle, B., Bell, S., Smeaton, A.F., McCusker, K., O'Connor, N.E.: Security considerations and key negotiation techniques for power constrained sensor networks. Comput. J. **49**(4), 443–453 (2006). https://doi.org/10.1093/comjnl/bxl023
9. Potdar, M.S., Manekar, A.S., Kadu, R.D.: Android 'Health-Dr.' application for synchronous information sharing. In: International Conference on Communication Systems and Network Technologies (CSNT), pp. 265–269. IEEE, Bhopal (2014). https://doi.org/10.1109/CSNT.2014.58
10. Wac, K., Bargh, M., Van Beijnum, B.J., Bults, R., Pawar, P., Peddemors, A.: Power- and delay-awareness of health telemonitoring services: the mobihealth system case study. IEEE J. Sel. Areas Commun. **27**(4), 525–536 (2009). https://doi.org/10.1109/JSAC.2009.090514

11. Halteren, V.: Mobile patient monitoring: the mobihealth system. J. Inf. Technol. Healthc. **2**, 365–373 (2009). https://doi.org/10.3233/978-1-60750-946-2-307
12. Roozbeh Jafari, A., Encarnacao, A.Z., Foad Dabiri, S. Noshadi Majid, H.: Wireless sensor networks for health monitoring. In: International Conference on Mobile and Ubiquitous System: networking and Services, pp. 479–481. IEEE (2005). https://doi.org/10.1109/MOBIQUITOUS.2005.65
13. Awasthi, A.K., Srivastava, K.: A biometric authentication scheme for telecare medicine information systems with nonce. J. Med. Syst. **37** (2013). https://doi.org/10.1007/s10916-013-9964-1
14. Du, X., Xiao, Y., Guizani, M., Chen, H.-H.: An effective key management scheme for heterogeneous sensor networks. Ad Hoc Netw. **5**(1), 24–34 (2007). https://doi.org/10.1016/j.adhoc.2006.05.012
15. Du, X., Xiao, Y., Ci, S., Guizani, M., Chen, H.-H.: A routing-driven key management scheme for heterogeneous sensor networks. In: 2007 IEEE International Conference on Communications, pp. 3407–3412 (2007). https://doi.org/10.1109/ICC.2007.564
16. Szczechowiak, P., Collier, M.: TinyIBE: identity-based encryption for heterogeneous sensor networks. In: 2009 5th International Conference on Intelligent Sensors, Sensor Networks and Information Processing (ISSNIP), pp. 349–354 (2009). https://doi.org/10.1109/ISSNIP.2009.5416743
17. Yussoff, Y.M., Hashim, H., Baba, M.D.: Identity-based trusted authentication in wireless sensor network. Int. J. Comput. Sci. Issues **9**(3) (2012)
18. Mishra, D.: On the security flaws in ID-based password authentication schemes for telecare medical information systems. J. Med. Syst. **39**(1) (2014). https://doi.org/10.1007/s10916-014-0154-6
19. Zhou, Y., Zhang, Y., Fang, Y.: Access control in wireless sensor networks. Ad Hoc Netw. **5**(1), 3–13 (2007). https://doi.org/10.1016/j.adhoc.2006.05.014
20. Tan, Z.: An efficient biometrics-based authentication scheme for telecare medicine information systems. Przegl. Elektrotech. **89**(5), 200–204 (2013). https://doi.org/10.1007/s10916-012-9856-9
21. Tan, Z.: A user anonymity preserving three-factor authentication scheme for telecare medicine information systems. J. Med. Syst. **38**(3) (2014). https://doi.org/10.1007/s10916-014-0016-2
22. Chatterjee, S., Das, A.K.: An effective ECC-based user access control scheme with attribute-based encryption for WSN. Secur. Commun. Netw. **9**(22), 5968–5974 (2014). https://doi.org/10.1002/sec.1140
23. Das, A.K.: Analysis and improvement on an efficient biometric-based remote user authentication scheme using smart cards. IET Inf. Secur. **5**(3), 145 (2011). https://doi.org/10.1049/iet-ifs.2010.0125
24. Lee, C.-C., Hsu, C.-W.: A secure biometric-based remote user authentication with key agreement scheme using extended chaotic maps. Nonlinear Dyn. **71**(1–2), 201–211 (2013). https://doi.org/10.1007/s11071-012-0652-3
25. Arshad, H., Nikooghadam, M.: Three-factor anonymous authentication and key agreement scheme for telecare medicine information systems. J. Med. Syst. **38**(12) (2014). https://doi.org/10.1007/s10916-014-0136-8

Strengthening E-Health Security: Authentication Analysis Through Formal Methods

Nazhatul Hafizah Kamarudin, Yusnani Mohd Yussoff, and Tutut Herawan

Abstract This chapter introduces and analyzes a secure, passwordless authentication framework for mobile e-health systems, with a focus on protecting sensitive patient data against common security threats. By moving away from traditional password-based methods, the research addresses vulnerabilities inherent in such systems, which are particularly challenging in the context of healthcare. The framework leverages unique identity-based encryption and a secure boot process that integrates Near Field Communication (NFC) technology to authenticate e-health sensor nodes, enhancing both security and user experience. The study utilizes formal methods, specifically the Automated Validation of Internet Security Protocols and Applications (AVISPA) tool, to validate the proposed protocol. This approach enables a thorough assessment of the protocol's robustness against attacks such as replay and node cloning. Through simulations, the protocol demonstrates resilience, confirming that unauthorized entities are effectively prevented from accessing the e-health network. The identity-based encryption scheme reduces memory consumption by up to 65%, while also optimizing energy usage, which is critical for resource-limited mobile e-health devices. Results from AVISPA simulations show that the authentication protocol can secure mobile e-health communications by safeguarding patient data without relying on third-party servers. The protocol's lightweight design minimizes computational overhead, making it suitable for real-time healthcare applications. Additionally, this chapter identifies areas for further research, such as refining the protocol for deployment in diverse healthcare settings and evaluating it in real-world environments. The findings contribute significantly to the advancement of

---

N. H. Kamarudin (✉)
Center for Cyber Security, Faculty of Information Science and Technology, Universiti Kebangsaan Malaysia, Bangi, Malaysia
e-mail: nazhatulhafizah@ukm.edu.my

Y. Mohd Yussoff
School of Electrical Engineering, College of Engineering, Universiti Teknologi MARA, Shah Alam, Malaysia

T. Herawan
Department of Information Systems, Faculty of Computer Science and Information Technology, University of Malaya, Kuala Lumpur, Malaysia

N. H. Kamarudin et al. (eds.), *Beyond Passwords: Protecting Healthcare with Authentication Security*, Information Systems Engineering and Management 51,
https://doi.org/10.1007/978-3-031-92247-3_6

secure mobile e-health systems, offering a practical solution that aligns with the demands of modern healthcare. This research lays a foundation for developing more sophisticated, resilient e-health authentication protocols, thereby supporting the growing adoption of digital healthcare technologies while prioritizing patient privacy and data security.

Keywords Mobile e-health security · Passwordless authentication · Identity-based encryption · Formal security analysis · AVISPA

1 Introduction

Authentication is the verification of the identity of an entity with a valid identity that has generated some data or with the integrity of the data [1]. In networking, authentication is a process in which the user establishes a right to access the resources. In recent security system, knowledge-based schemes are predominantly applied for user authentication which demands a high level of memorability from the user such as login name and password. However, username or login name and passwords have a number of shortcomings despite being simple and are easier to remember. One being its vulnerability to security attacks. While passwords that are complex and arbitrary provide more security, they are more difficult to remember. Password-authentication does not provide an adequate mechanism for secure authentication since they may be forgotten or stolen. Moreover, it cannot prove the actual existence of the user. The password is transmitted from the user into the network system in clear text on communication line which can expose the rogue system to abstract the password. Even if the passwords are encrypted or hashed, there is often no restriction on reading the password file using brute force cracking techniques or any other methods.

With rapid technological advancement on the security features, it demands mobile e-health network to increase its security level to protect its private credentials and sensitive information from being exposed or manipulated by the other parties [2, 3]. This thesis proposes a new secure authentication procedure through a complete framework from the establishment of the unique identity of the e-health sensor node to the passwordless authentication protocol, and finally reports or transmits the message to the trusted authority which is the e-health server.

2 Passwordless Authentication in E-Health

The design of the framework with the development of passwordless authentication using a unique identity based encryption is quite a difficult and challenging task that requires the thorough understanding of the e-health network architecture concept with the details exploration on the e-health sensor node identity. This thesis presents a detailed conceptual design on the framework of mobile e-health network architecture with the consideration of the energy limitation by the e-health sensor nodes as well as the mobile device. A new method for extracting the e-health sensor node identity features is presented and the design of the secure boot process provides a secure method to validate the e-health sensor node in the mobile e-health network.. The unique identity of the e-health sensor node is a combination of certain features which are the secure boot identity and the NFC ring identity. This research contributes in solving the issues related to sensor node cloning or node impersonation from the lowest level during a boot-up stage in a secure boot process. Even if a duplicated node may be able to intercept the session identity transmitted over the network, it will definitely fail to authenticate with the e-health server due to the inexistence of a valid and unique value in the e-health sensor node trust list in the e-health server. A statistical approach is used to calculate the energy consumptions as well as the security verification of the proposed protocol. Proposed lightweight IBE_EHEART algorithm has successfully reduced up to 65% of memory storage consumption in the mobile e-health system. Evaluation results obtained for the energy consumptions on selected major processes in the proposed authentication framework are found to be lower than the other researches that utilize password based authentication protocol.

Another contribution is providing an in-depth research on mobile e-health network architecture and authentication protocol for a secure communication transmission in a mobile e-health system. The research in this area resulted in the development of a new passwordless authentication protocol for the e-health sensor node and mobile device in mobile e-health network. Moreover, it contributes to significant savings on cryptographic keys storage and energy consumption while increasing the feasibility of the mobile e-health application. The proposed framework enables systematic procedures in differentiating trusted and non-trusted e-health sensor node in the mobile e-health network. Through this proposed approach, the e-health sensor node authenticity is no longer depending on the e-health third party service provider or a knowledge-based evaluation scheme. This protocol enables to protect user anonymity and mitigate the replay attack that takes place in the mobile e-health network. Finally, a new secure passwordless authentication protocol in mobile e-health which is proven to be secured from the replay attack and sensor node cloning attack through AVISPA model checking. This proposed authentication protocol has also contributed to the framework development of a two-tier mobile e-health network architecture which is able to directly communicate with the e-health server without any involvement from the third party server.

3 Security Analysis

3.1 Analysis on Formal Method Security Features

Formal methods analysis is a mathematical-based technique for the software and hardware system verification. The specifications are used as the basis of proving properties to produce correctness of a system. A specification is a model of the designed system or protocol that contains a description of its desired behaviour that will be implemented throughout the system [4]. The verification of the designed system or protocol specifications is needed by presenting a formal analysis proof. There are several automated means such as automated theorem proving and model checking that can produce formal analysis proof of a system specification. Theorem proving has to produce a formal proof from a scratch with a description of the system together with a set of logical axioms and a set of inference rules. On the other hand, model checking is a verification tool that can verify certain properties by means of exhaustive search of all possible states during the system execution. It is more practical to verify a system protocol using a model checking tool since it is an automated technique for formal software verification which is not restricted to input processing output paradigm and it also can be applied to a system with realistic properties [5]. Several examples of significant model checking tools have been discussed thoroughly in Chap. "Authentication Strategies for Modern Healthcare Systems". This research has chosen AVISPA as the model checking tool for the formal method security analysis.

The AVISPA is a push-button tool for automated validation of Internet security sensitive protocols and applications which formally verifies whether a security protocol is safe or unsafe. It integrates different back-ends that implement a variety of state-of-the-art automatic analysis techniques. There are for back-ends in AVISPA tool analysis which are On-the-fly Model-checker (OFMC), Constraint Logic based Attack Searcher (CL-AtSe), SAT-based Model-Checker (SATMC), and finally Tree Automata based on Automatic Approximations for the Analysis of Security Protocols (TA4SP). In AVISPA, the protocols need to be specified in High Level Protocols Specification Language (HLPSL) that is a role-oriented language and each specified role is independent from the other roles. In the role section, the system defines the number of sessions and principals. Additionally, an intruder is modelled using the Dolev-Yao model with the possibility of a legitimate role for the intruder in a protocol simulation run. The HLPSL code is then converted into an intermediate format (IF) using the HLPSL2F translator.

The output format of AVISPA is generated through that four back-ends tools and there are several output sections which is first the SUMMARY section to indicate the security status of the tested protocol either it is safe or unsafe. Next section is DETAILS where the explanation of the tested protocol is given and why it is declared as safe or under what conditions or attacks that makes the protocol is unsafe. Other sections which are PROTOCOL, GOAL, and BACKEND represent the name of the tested protocol, the goal of the security analysis and which back-end tool that is used in order to verify the security protocol. Last but not least, there are some comments and statistics of the tested protocols as well as any trace of an attack if any are also displayed in the standard Alice-Bob format. More details on AVISPA have been discussed previously in Sect. "Formal Security Analysis".

A lot of researches nowadays have implemented AVISPA as their formal analysis method in order to analyse their newly designed security protocols. Odelu et al. [6–8] has introduced a secure authentication protocol for NFC application in its mobile healthcare system to overcome the impersonation attacks as well as man-in-the-middle attacks. Their authentication protocol is simulated using AVISPA tool to show that the protocol is secure. The simulation results for the formal security verification are done using the OFMC and CL-AtSe backends and the authentication protocol is proven to be secure against replay and man-in-the-middle attacks.

Das et al. [9, 10] implemented a biometric authentication scheme for the mobile e-health system based on three factor remote user authentication scheme. Das authentication scheme is an enhancement from the Awasthi-Srivastava's scheme [11] which has failed to protect intense replay attack and failed to preserve user anonymity besides the scheme lacks of rigorous formal security analysis. Das et al., then simulate their scheme for the formal security verification using AVISPA to show that their authentication scheme is secure against replay attack. The authentication scheme is specified thoroughly based on the role specification in HLPSL language. The role of intruder also participates in the protocol execution in the AVISPA analysis. Das et al., has successfully simulated their scheme and ensure that it is secure against replay attack. However, it does not prove any user anonymity properties in their security analysis.

Tan authentication scheme [12] in mobile e-health is developed to provide the user anonymity property in order to prevent the compromise of the real identity of a user. However, a formal security analysis has been done towards Tan's authentication scheme and the result clearly shows that Tan's scheme is vulnerable against replay attacks. The security analysis summary indicates that the protocol is unsafe since attacks are found in Tan's authentication scheme.

Chatterjee et al. [13] have also used AVISPA tool for their formal security verification in their proposed mobile e-health authentication protocol based on attribute-based encryption and elliptic curve cryptography. Chatterjee et al., have implemented three basic roles in the HLPSL language which are alice, bob, and bs. Their authentication protocol have also verified using the OFMC backend and it is verified to be secured from the replay attack. Amin et al. [14] have also use AVISPA in their user authentication protocol verification in telecare medical information system. The formal security analysis is presented as a proof that the proposed authentication protocol is secure and safe against the adversaries. They have presented the roles for each entity in their authentication protocol for providing registration to the new user and medical server in HLPSL language. Based on the Amin et al., AVISPA simulation results, their authentication protocol is shown to be secure from replay attacks.

Many researches have been done to enhance the security of the mobile e-health authentication scheme and most of them are using formal analysis to verify the protocol security analysis. AVISPA is one of the formal analysis verification tools that have been widely accepted and thus it is chosen as the security analysis verification tool for the EHEART authentication protocol. The details on AVISPA security analysis on the EHEART authentication scheme will be discussed in the next subsection.

3.2 Formal Analysis on EHEART Authentication Protocol

The security of the EHEART protocol is best realized through the security of full Boneh Franklin Identity-Based Encryption scheme. In this scheme, the public key of the BS is generated from some random value and thus to be used in the generation of ciphertext or encrypted login information. EHEART authentication protocol is simulated using the widely-accepted AVISPA which is a push-button tool for automated validation of Internet security sensitive protocols and applications. It will then formally verify whether the security protocol is safe or unsafe. Additionally, the simulation results using AVISPA clearly confirm that the proposed authentication protocol is secure against security attacks which are replay attack and sensor node cloning attack as well as user anonymity protection.

Formal analysis on the proposed authentication protocol is conducted to validate the security features of the authentication scheme and to confirm its functionality. Dolev-Yao threat model [15] is used in the proposed authentication protocol where the communication between e-health sensor node and mobile device as well as the e-health server is transmitted over a public channel which is considered as insecure. The e-health server is assumed to be trustworthy and the attacker is not be able to compromise with the server. Throughout the security analysis, several assumptions are put into the consideration. First, the public key and common security parameters have been installed in the e-health sensor node and mobile device. Second, the e-health server or the e-health base station has already been deployed with the list of a valid e-health sensor node identity as well as its unique value generated after the secure boot process. For the purpose of the security analysis, the network protocol model is simulated on the e-health sensor node, mobile device, and the e-health server.

The proposed authentication scheme is implemented in the form of HLPSL language where three different network scenarios are formally analysed and thoroughly discussed in the subsequent paragraphs.

a. **Scenario 1**

In this scenario, $IDSN_A$, identity of a valid e-health sensor node attempts to join the e-health network and demands to transmit the e-health data. Its sensor node identity is listed in *esnlist* in the e-health server or BS.

Upon registration, e-health sensor node A ($IDSN_A$) will go to a secure boot process and produce a non-regenerated unique identity value, Sb_A'. E-health sensor node A ($IDSN_A$) sends encrypted login information containing Sb_A' to the mobile device

(M) for authentication. Mobile device (M) will verify the e-health sensor node identity in the *esnlist* located in the e-health server (BS). E-health server will then check the identity of the mobile device using the NFC identity as well as the identity of the e-health sensor node A and proceeds with the authentication process if the identity exists in the *esnlist.* The authentication process is simulated in HLPSL language as shown in Figs. 1 and 2.

```
% SN--> M.  SN send login information to M.
% M checks IDSN in the e-health server BS.
% BS checks the M identity. BS replies and confirms that IDSN is valid thus M can decrypt the login information
%--------------------------------

role sensor (SN, M       : agent,
             IDSN,IDM    :text,
             Mac,Hash    : hash_func,
             Ksn         :public_key,
             Snd,Rcv     :channel(dy))

played_by SN def=

local   State   : nat,
        Nsn     : text,
        X       : text,     %random number
        W       : hash(text.public_key.text.text),   %H(X1,IDSN|IDM) locally calculate as session key
        X1      : text.public_key,  %X1=X*Ksn  to be send to mobile
        Ksnm     : symmetric_key,   %e((X1.W)Ssn,Km) where Ssn=inv(Ksn)
        Km      :public_key

init State :=0

transition

1.State = 0 /\ Rcv(start) =|>
  State':= 1 /\ Nsn':= new()
             /\ X' := new()
             /\ X1':= X'.Ksn
             /\ W' := Hash(X1'.IDSN.IDM)        %idSN is a non-regenerated sensor node unique identity from secure boot process
             /\ Ksnm':= exp((inv(Ksn).X'.W),Km)    %locally calculate symmetric key
             /\ Snd(SN.M.IDSN.X1'.Nsn'.Mac(IDSN.X1'.Nsn'))
                     /\ secret(Ksnm,sym_ksn,{SN,M})

end role
```

Fig. 1 E-health sensor node role in EHEART Authentication process

```
role mobile (SN,M                  :agent,
          IDSN,IDM        :text,
          Hash,Mac        :hash_func,
          Snd,Rcv:channel(dy))

played_by M def=
local             State : nat,
                  TL      :(text) set,
                  Nsn     : text,
                  Kmsn    : symmetric_key  ,   %e((X1+W.inv(Ksn),Km)
                  W       : hash(text.public_key.text.text),   %H2(X1,IDSN|IDM) locally calculate as symmetric key
                  X1      : text.public_key,  %X1=X*Ksn   to be send to M
                  Km,Ksn  :public_key

init State := 2

transition

1.State = 2 /\ Rcv(SN.M.IDSN.X1'.Nsn'.Mac(IDSN.X1'.Nsn'))  =|> %/\in(IDSN,TL)=|>  %check in BS,receive login information from IDSN
State' := 4 /\ W':= Hash(X1'.IDSN.IDM)
           /\ Kmsn' := exp((Ksn.X1'.W'),inv(Km))  %locally calculate Kmsn
           /\ secret(Kmsn,sym_km,{M,SN})

%2.State = 4 /\ Rcv(idSN.X1'.Nsn'.Mac(IDSN.X1'.Nsn')) /\ not(in(idSN,TL)) =|>     %receive message from BS, IDSN is valid
%State' :=6    %if not in TL M will not proceed

end role
```

Fig. 2 E-health mobile device role in EHEART authentication process

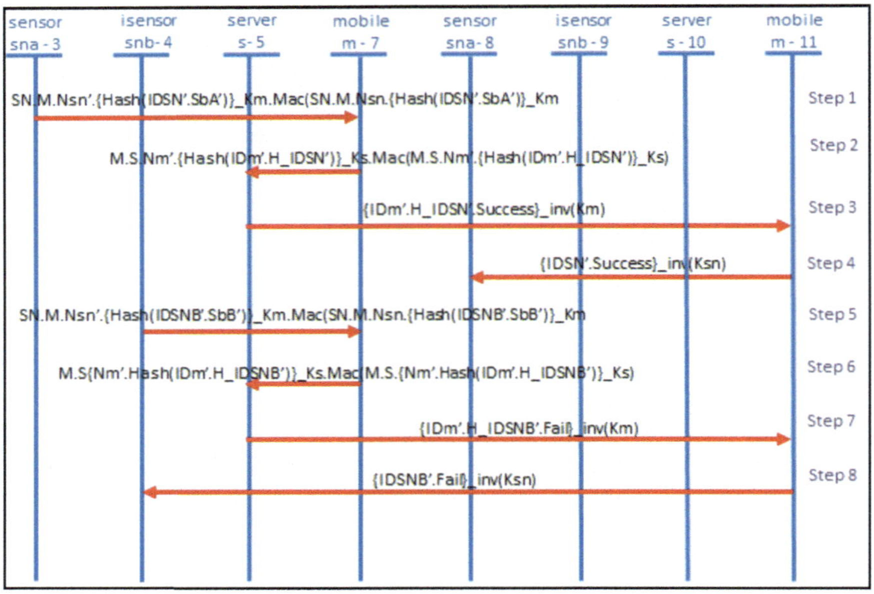

Fig. 3 Simulated version of a secure mobile e-health authentication scheme

Upon successful verification, e-health server then sends acknowledgement message or packet to the e-health sensor node through the mobile device as a communication gateway and it allows the mobile device to receive login information from the e-health sensor node A otherwise the e-health server replies with a 'FAIL' message to the invalid e-health sensor node. It can be referred to Fig. 3.

Mobile device will then approve the login attempt from the e-health sensor node and receive the data and store it in the e-health server. Figure 4 shows the details on the authentication between the mobile device and the e-health server where the BS needs to verify the identity of the mobile device first before receiving the login information from the e-health sensor node that has been transmitted through the mobile device (Fig. 5).

b. **Scenario 2**

In this scenario, $IDSN_B$, an illegal e-health sensor node attempts to join the e-health network. However, its identity is not listed in *esnlist* located in the e-health server database (BS).

Consider a situation where an intruder with invalid identity trying to join into the e-health network. The mobile device will transmit the login information which contains the received IDSN to BS after NFC ring identity configuration. The BS then compares the received IDSN with the list of e-health sensor node identity in *esnlist*. At this particular phase, assume only $IDSN_A$ exists in *esnlist*. For that reason, BS will send a reply message to M with failure notification. This communication can

```
role mobile (M, S: agent,
            Km,Ks:public_key,  %Km,Ks = H(IDX)
            IDm: text,
           KeyRing:(agent.text) set,
            Snd,Rcv:channel(dy))

played_by M def=

local State : nat,
Nm, Nsn: text


%Ksn:public_key

init State := 0

transition
% Compare IDm value in mobile identity with e-health base station

1. State = 0 /\ Rcv(start) =|>
   State':= 2 /\ IDm' := new() %assume the IDm is just generated by the mobile device which is using the NFC ring ID
              /\ Nm' := new() /\ Snd(S.M.{M.IDm'.Nm'}_Ks) %Send IDm which is the ID mobile and send to e-health server   %
              %/\ secret(Nm',snm,{M,S})
              /\ secret(IDm',sidm,{M,S})
              %/\ witness(M,S,mobile_server_nm, Nm')
             % /\ witness(M,S,mobile_server_idm, IDm')


2.State = 2 /\ Rcv(M.S.{Nm.Nsn'}_Km) =|>
  State':= 4 /\ Snd(S.M.{Nsn'}_Ks)
            /\ request (M,S,mobile_server_nm, Nm)
             /\ request(S,M, mobile_server_idm, IDm) %authenticate using nonce mobile nm' and nonce sensor node nsn'. Then, auntenticate on IDm
              /\ Rcv(tn)

end role
```

Fig. 4 Mobile device authentication by the e-health server

```
role server ( S:agent,
            Ks,Km:public_key,
            KeyMap:(agent.text) set,

Snd,Rcv: channel(dy))

played_by S def=

local State : nat,
     M :agent,
Nsn, Nm: text,
%TrustMobile:agent list,
     IDm:text

init State := 1
transition
1. State = 1 /\ Rcv(S.M.{M.IDm'.Nm'}_Ks) /\ in(M.IDm, KeyMap) =|>
   State' := 3  /\ Nsn':= new()
                /\ Snd (M.S.{Nm'.Nsn'}_Km)



2. State = 3 /\ Rcv(S.M.{Nm}_Ks) =|>
   State' := 5  /\request(S,M,mobile_server_nsn,Nsn) %agent S accept the value Nm' and now relies on agent M to agree on the value Nsn'/\      Snd(tn)

end role
```

Fig. 5 E-health server role specification in authenticating the mobile device

also be referred in Step 5 until Step 8 in Fig. 12 in Chap. "Performance Analysis of Mobile E-Health Authentication System". $IDSN_B$ subsequently will be blacklisted from the *esnlist* and will no longer be able to transmit the data. BS will not even encrypt the message due to the identity authentication failure of $IDSN_B$.

c. **Scenario 3**

In this scenario, a rogue sensor node has stolen the e-health sensor node A identity, $IDSN_A$ and attempts to join the e-health network with an invalid secure boot non-regenerated unique identity value, Sb_A.

This situation happens when the perpetrator knows the e-health sensor node session identity used in the e-health network. The perpetrator might capture the e-health sensor node session identity and attempts for sensor node cloning by using the same $IDSN_A$ to get authenticated. Upon successful IDSN verification, BS will decrypt the login information and compare the secure boot non-regenerated unique identity value, Sb_A. However, BS will detect a different Sb_A value since the value is unique and it needs to be generated through a secure boot process and the Sb_A value cannot be cloned by the intruders because of its unique trust measurement value. Thus, due to a contraction between the e-health sensor node identity and the Sb_A value, BS identifies the intruder and sends a failure reply notification to M.

3.3 EHEART Protocol Specifications

The three scenarios are simulated using the HLPSL language. The security of the EHEART protocol is based on asymmetric cryptography where the login information is encrypted with the Ks which is the e-health server (S) public key and it is derived from e-health server identity. In the implementation of EHEART in HLPSL for the session key agreement phase, *sn*, *m*, and *ss* are the three basic roles which represent the participants of the EHEART system which are the e-health sensor node, mobile device and the e-health server.

The communication is established between the e-health sensor node A known as *sn* and e-health server known as *s* through a mobile device gateway known as *m*.

Referring to Table 1, a secure communication has been established between the e-health sensor node known as *sn*, a mobile device known as *m*, and e-health server known as *bs*. The AVISPA commands represent the packet sent and received as well as the communication between *sn*, *m* and *bs*. The prime "'" symbol used in the commands protocol indicate the fresh value generated from the entities. Since the non-regenerated unique identity of the e-health sensor node *IDSN* and the mobile device identity *IDm* which are the NFC_{ID} are secure values that must be protected and known only to the e-health server, AVISPA formal analysis will confirm its confidentiality through the command protocol *secret(IDSN',sidsn,{SN,M})* and *secret(IDm,sidm,{M,S})*. This AVISPA commands will ensure the integrity and the confidentiality of *IDSN* and *IDm* with the purpose of only *sn*, *m* and *bs* know those values. The *sidsn* and *sidm* is the identifier syntax in the function *secret*. Besides, the authenticity of the e-health sensor node and also the mobile device identity which is the NFC identity are established through the command *request* and *witness*.

During the *sn* and *m* communication channel, the e-health sensor node *sn* requests to be authenticated based on the *IDSN'* value through the AVISPA command protocol *request(M,SN,sensor_mobile_idsn,IDSN')* and *witness(SN,M,sensor_mobile_idsn,IDSN')*. On the other hand, during the *m* and *bs* communication channel, the mobile device *m* requests to be authenticated based on the *IDm'* value through the AVISPA command protocol *request(S,M,mobile_server_idm,IDm')* and *witness(M,S,mobile_server_idm,IDm')*. The *sensor_mobile_idsn* and *mobile_server_idm* are defined as the protocol ID in the HLPSL coding.

Table 1 EHEART authentication commands

EHEART participants	AVISPA commands
sn → m	*Snd(SN.M.Nsn′.{Hash(IDSN'.SbA')}_ Km.Mac(SN.M.Nsn'.{Hash(IDSN'.SbA')}_Km))* *∧secret(IDSN′,sidsn,{SN,M})* *∧witness(SN,M,sensor_mobile_idsn,IDSN′)*
m → sn	*Rcv(SN.M.Nsn'.{Hash(IDSN')}_ Km.Mac(SN.M.Nsn'.{Hash(IDSN)′}_Km))* *∧in(IDSN,TL)=|>* *∧request(M,SN,sensor_mobile_idsn,IDSN')* *∧ Snd({IDSN.Success}_inv(Ksn))*
m → s	*Snd(M.S.Nm′.{Hash(IDm′.H_IDSN')}_ Ks).Mac(M.S.Nm′.{Hash(IDm′.H_IDSN')} _Ks))* *∧ secret(IDm′,sidm,{M,S})* *∧witness(M,S,mobile_server_idm, IDm′)*
s → m	*Rcv(S.M.Nm′.{Hash(IDm′.H_IDSN')}_ Ks.Mac(S.M.Nm′.{Hash(IDm'.H_IDSN′)}_Ks))* *∧ in(M.IDm, KeyMap)=|>* *∧ request(S,M,mobile_server_idm,IDm')* *∧ Snd ({IDm'.H_IDSN'.Success}_inv(Km))*

In addition, to secure the *IDSN* and *IDM* values, the protocol sends the hash values instead of the exact values so that even if the adversary manages to acquire the login information and decrypt the message, the adversary still fails to get the exact identity value but only the hash management value.

However, the authentication protocol is still UNSAFE in both Cl-Atse and OFMC backends model where the "*authentication_of_idsn*" and "*authentication_of_idm*" have been violated. In the next subsection (a) and (b), the vulnerabilities found are discussed thoroughly and the improvement made to eliminate the attacks is presented.

a. **Attack analysis**

From the attack trace presented in Fig. 15 in Chap. "Performance Analysis of Mobile E-Health Authentication System", the first legitimate session begins with the e-health sensor node A "*(a − 3)*" or *sna*, sending the first message to mobile device, *M* or *sm*. Then the mobile device will send the message to the e-health server, *ss*. However, it is being intercepted by an intruder (*i*). The intruder then forward the message to e-health server (BS), "*(s, 4)*" or *s* (Referred to Step 3 and Step 4 in Fig. 15 in Chap. "Performance Analysis of Mobile E-Health Authentication System"). Following that, e-health server, *s* reply the next message of the protocol to intruder "*i*" instead of the mobile device, *m*. The intruder then simply forwards the message to the mobile device, *m*. Then, the mobile device sends the message to the e-health sensor node A. At this moment, the intruder is just doing the passive attack by eavesdropping to the message and silently forwards the message. However, further analysis using the Security Protocol Animator (SPAN) has shown that the intruder can play an active role after eavesdropping the message by sending the second message of the authentication protocol to the mobile device without any respond from the e-health server. The subsequent message of the protocol is forwarded without the participation of the e-health server and it means that the intruder has taken over control of the e-health server, S or *s*. Referring to Fig. 6, step 8 where *mobile(m-9)* transmit the protocol message to *intruder(i-0)* which supposed to be *server(s)* and step 9 where *intruder(i-0)* replies back to *mobile(m-9)* proves the statement that the subsequent protocol messages sent by e-health sensor node and mobile device are intercepted by the intruder and forwarded back to respective mobile device without the e-health server participation at all. Thus, the intruder has taken control over the e-health server.

b. **Protocol Improvement**

To overcome this replay attack, a few modifications on the authentication have been done as shown in Fig. 7. To avoid the intruder from replying similar messages to the mobile device and e-health sensor nodes, nonce values *Ns'* and *Nm'* are included in the reply messages. Thus, the receiver which is the mobile device as well as the e-health sensor node will authenticate the reply message to validate the message freshness.

The e-health server will also verify the validity of the *IDm'* in the *KeyMap* list by decryption of the packet and comparing the hash value of *H_IDSN'* which contains the hash value of the e-health sensor node identity and the e-health sensor node secure boot value with the *esnlist*. The e-health server will successfully authenticate

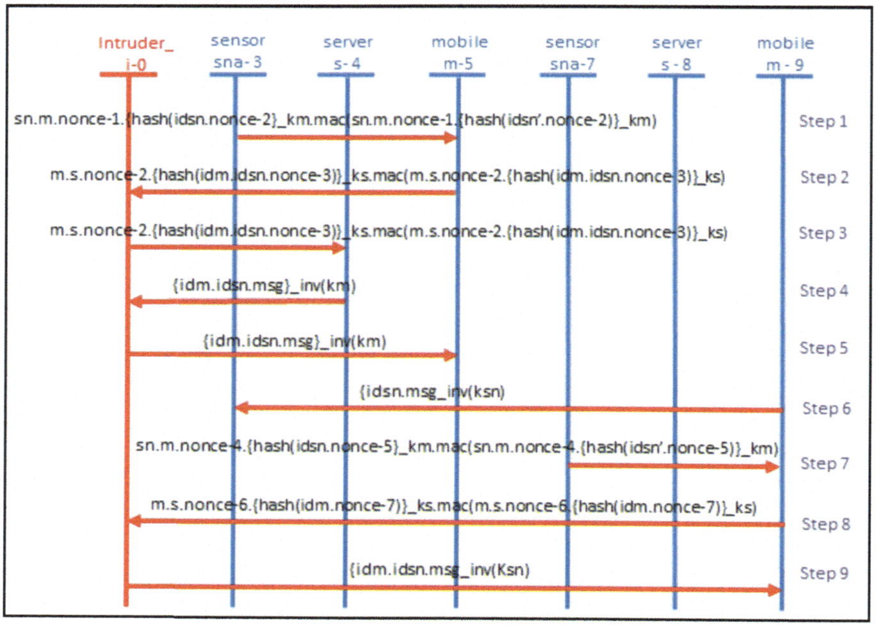

Fig. 6 Attack analysis on the proposed mobile e-health authentication

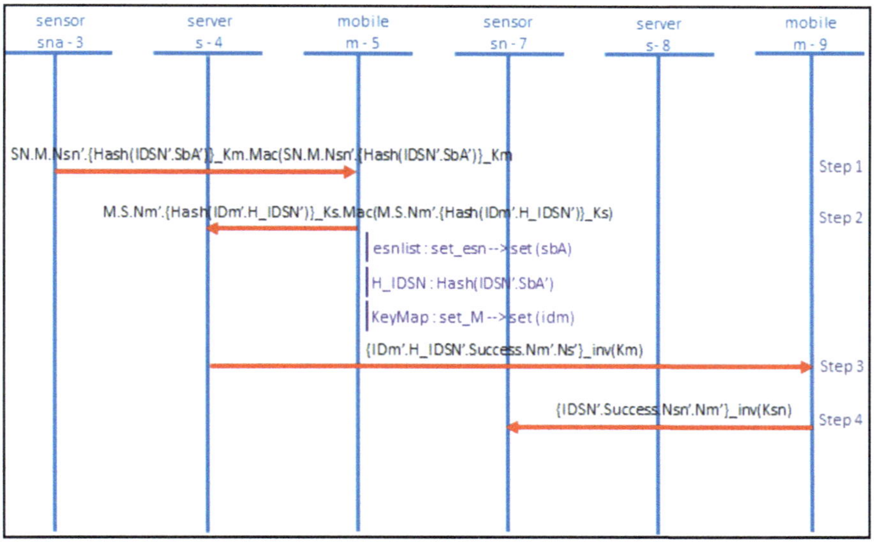

Fig. 7 Simulated version of the improved mobile e-health authentication scheme

the e-health sensor node if only both of the values *IDm'* and *H_IDSN'* are equal. Thus, based on the following modification presented in Table 2, OFMC and Cl-Atse have confirmed that the proposed authentication protocols are secure and safe.

In the new modified AVISPA protocol statements, the e-health server will verify the validity of the *IDm'* plus the *H_IDSN'* values and also the nonces are added in the reply messages as well so that the receiver can confirm the freshness of the message packet (Fig. 8).

The modified authentication protocol has eliminated the attack problems described before in subsection (a) where the intruder impersonates as the e-health server and intercept the messages from the mobile device. By the modification in the authentication protocol and the addition of the nonce values in the reply message, the intruder

Table 2 Modified protocol for security improvement

Mobile e-health participants	AVISPA commands
server	*Rcv(M.S.Nm'.{Hash(IDm'.H_IDSN')}_Ks.Mac(M.S.Nm'.{Hash(IDm'.H_IDSN')}_Ks))* *∧ in(IDSN,esnlist)* *∧ in(IDm,KeyMap)*
server → mobile	*Snd({IDm'.H_IDSN'.Success.Nm'.Ns'}_inv(Km))*
mobile	*Rcv(SN.M.Nsn'.{Hash(IDSN'.SbA')}_Km.Mac(SN.M.Nsn'.{Hash(IDSN'.SbA')}_Km))*
mobile → e-health sensor node	*Snd ({IDSN'.Success.Nsn'.Nm'}_inv(Ksn))*

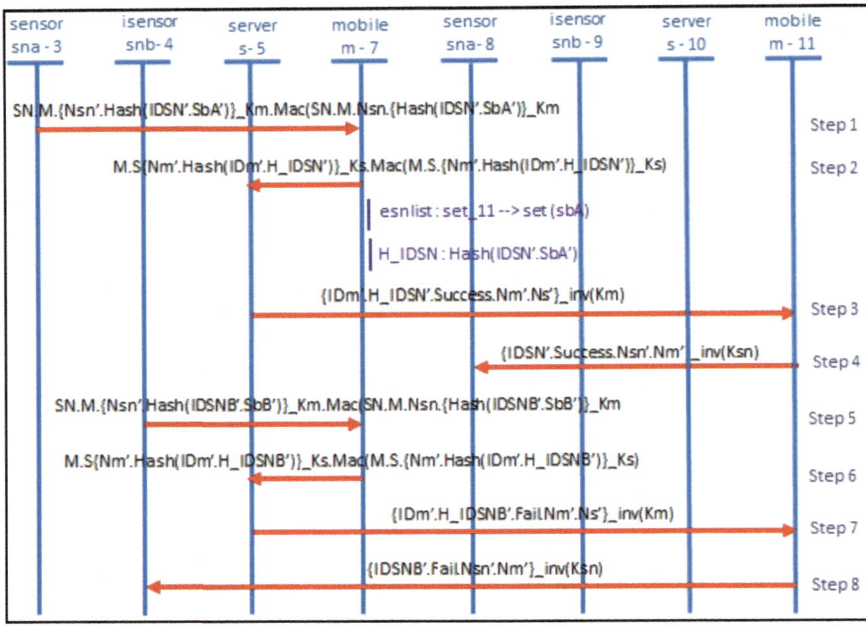

Fig. 8 Simulated full version of the improved mobile e-health authentication scheme

is unable to reply messages sent by the mobile device and the e-health server (Figs. 9, 10 and 11).

4 Future Works on Mobile E-Health Authentication

Ensuring secure communications between the e-health sensor node and mobile device with the e-health server is definitely one of the biggest challenges in providing a secure and feasible infrastructure for mobile e-health applications. This thesis presents novel methods in securing the e-health sensor network within a two-tier mobile e-health architecture. Nevertheless, there are more interesting research problems in this area that still require further investigations and researches that have not yet been fully addressed.

The analysis presented in the development of the unique identity of the e-health sensor node can be regarded as a pioneering work in establishing a unique identity of the e-health sensor node. Future work is needed to enable a complete analysis on the feasibility and the seamless application in mobile e-health system. Plus, the authentication protocol is developed within the laboratory area which includes the e-health sensor nodes, mobile device, and a server that acts as an e-health server.

```
% OFMC
% Version of 2006/02/13
SUMMARY
  SAFE
DETAILS
  BOUNDED_NUMBER_OF_SESSION
PROTOCOL
  C:\SPAN\testsuite\results\SN_Serve
GOAL
  as_specified
BACKEND
  OFMC
COMMENTS
STATISTICS
  parseTime: 0.00s
  searchTime: 0.20s
  visitedNodes: 328 nodes
  depth: 15 plies
```

```
SUMMARY
  SAFE

DETAILS
  BOUNDED_NUMBER_OF_SESSION
  TYPED_MODEL
  BOUNDED_SEARCH_DEPTH

PROTOCOL
  C:\SPAN\testsuite\results\SN_Server

GOAL
  As Specified

BACKEND
  CL-AtSe

STATISTICS

  Analysed    : 0 states
  Reachable   : 0 states
  Translation: 0.03 seconds
```

Fig. 9 OFMC and CL-AtSe backends on e-health sensor node and e-health server transmission

```
% OFMC
% Version of 2006/02/13
SUMMARY
  SAFE
DETAILS
  BOUNDED_NUMBER_OF_SESSION
PROTOCOL
  C:\SPAN\testsuite\results\SNtoMobil
GOAL
  as_specified
BACKEND
  OFMC
COMMENTS
STATISTICS
  parseTime: 0.00s
  searchTime: 0.01s
  visitedNodes: 8 nodes
  depth: 3 plies
```

```
SUMMARY
  SAFE

DETAILS
  BOUNDED_NUMBER_OF_SESSION
  TYPED_MODEL

PROTOCOL
  C:\SPAN\testsuite\results\SNtoMobil

GOAL
  As Specified

BACKEND
  CL-AtSe
```

Fig. 10 OFMC and CL-Atse backends on e-health sensor node and mobile device communication

Further research can be conducted in a real environment in the medical application to refine the assumptions made in designing the protocol.

Another future research that can be done is throughout the security analysis on the authentication protocol. The mobile e-health authentication protocol was formally

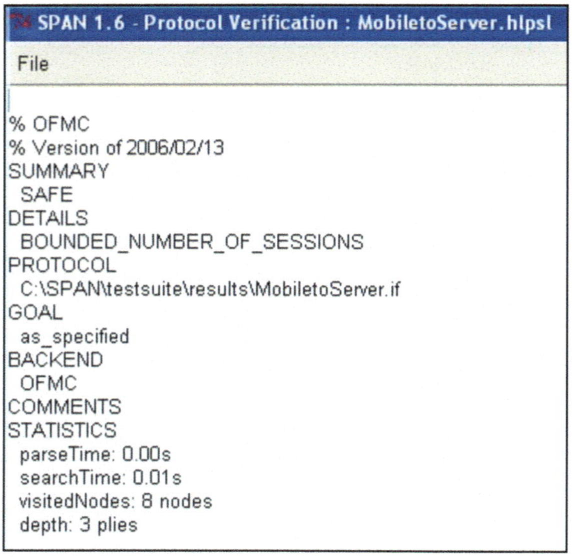

Fig. 11 Mobile device to e-health server OFMC backend

analysed on AVISPA and SPAN attack model. Thus, although the protocol has been formally analysed to be secure and safe from the security attacks, research on the security analysis can be further implemented and tested in a real environment to practically prove its security features.

5 Conclusion

Recent advancement in wireless technology system has enabled rapid development on the mobile e-health system. Higher demand on the mobile e-health system together with highly valuable data it can provide might expose the mobile e-health system to various security attacks. To ensure data security and data integrity, passwordless authentication for mobile e-health monitoring system could replace current health monitoring technology due to its low power consumption and reducing elders discomfort with its ease-of-use. Healthcare provider can also reduce its expenses on a third party server which can introduce to unforeseen vulnerability. Hence, health communication nowadays will become gradually easier and more practical due to the enhancement in mobile e-health application [16]. In the mobile e-health application, the security attack specifically on the e-health sensor node has been an emerging security issue of the network system. With the deployment of the e-health sensor node that is always been left unattended by the healthcare personnel has exposed the network to several potential vulnerabilities. Therefore, it is more than necessary to implement additional security features on this e-health sensor network platform so that it has the security requirement before it can be applied in the mobile e-health system [17–20]. This book explores the method for establishing unique identities and presents a passwordless authentication protocol. These innovations demonstrate the feasibility of integrating robust security measures into future mobile e-health networks, ensuring seamless authentication processes.

References

1. Van der Linden, H., Kalra, D., Hasman, A., Talmon, J.: Inter-organizational future proof EHR systems. A review of the security and privacy related issues. Elsevier Int. J. Med. Inform. **78**(3), 141–160 (2009). https://doi.org/10.1016/j.ijmedinf.2008.06.013
2. Rajeshkumar, K., Dhanasekaran, S., Vasudevan, V.: A novel three-factor authentication and optimal mapreduce frameworks for secure medical big data transmission over the cloud with shaxecc. Multimed. Tools Appl. (2024). https://doi.org/10.1007/s11042-024-18147-6
3. Saranya, A., Naresh, R., Karuppiah, S., Jenifer, M.: Development of trust-based authorization and authentication framework for secure electronic health payment in cloud environment. Soft Comput. (2024). https://doi.org/10.1007/s00500-023-09514-w
4. Almeida, J.B., Frade, M.J., Pinto, J.S., Melo de Sousa, S.: Rigorous software development (2011). https://doi.org/10.1007/978-0-85729-018-2
5. Halpern, J., Vardi, M.: Model checking versus theorem proving: a manifesto. In: Artificial Intelligence and Mathematical ..., pp. 1–30 (1991). https://doi.org/10.1016/B978-0-12-450010-5.50015-3
6. Odelu, V., Das, A.K., Goswami, A.: SEAP: secure and efficient authentication protocol for NFC applications using pseudonyms. IEEE Trans. Consum. Electron. **62**(1), 30–38 (2016). https://doi.org/10.1109/TCE.2016.7448560
7. Odelu, V., Das, A.K., Goswami, A.: An efficient key-management scheme for hierarchical access control in e-medicine system. J. Med. Syst. **36**(4), 2325–2337 (2013). https://doi.org/10.1007/s10916-011-9700-7
8. Odelu, V., Das, A.K., Khurram Khan, M., Choo, K.K.R., Jo, M.: Expressive CP-ABE scheme for mobile devices in IoT satisfying constant-size keys and ciphertexts. IEEE Access**5**, 3273–3283 (2017). https://doi.org/10.1109/ACCESS.2017.2669940
9. Das, A.K.: A secure user anonymity-preserving three-factor remote user authentication scheme for the telecare medicine information systems. J. Med. Syst. **39**(3) (2015). https://doi.org/10.1007/s10916-015-0218-2
10. Das, A.K., Goswami, A.: An enhanced biometric authentication scheme for telecare medicine information systems with nonce using chaotic hash function. J. Med. Syst. **38**(6), 27 (2014). https://doi.org/10.1007/s10916-014-0027-z
11. Awasthi, A.K., Srivastava, K.: A biometric authentication scheme for telecare medicine information systems with nonce. J. Med. Syst. 37 (2013) https://doi.org/10.1007/s10916-013-9964-1
12. Tan, Z.: A user anonymity preserving three-factor authentication scheme for telecare medicine information systems. J. Med. Syst. **38**(3) (2014). https://doi.org/10.1007/s10916-014-0016-2
13. Chatterjee, S., Das, A.K.: An effective ECC-based user access control scheme with attribute-based encryption for WSN. Secur. Commun. Netw. **9**(22), 5968–5974 (2014). https://doi.org/10.1002/sec.1140
14. Amin, R., Biswas, G.P.: A novel user authentication and key agreement protocol for accessing multi-medical server usable in TMIS. J. Med. Syst. **39**(3) (2015). https://doi.org/10.1007/s10916-015-0217-3
15. Dolev, D.: On the security of public key protocols. IEEE Trans. Inf. Theory **29**(2), 198–208 (1983). https://doi.org/10.1109/SFCS.1981.32

16. Bathalapalli, V.K.V.V., Mohanty, S.P., Kougianos, E., Iyer, V., Rout, B.: PUFchain 3.0: hardware-assisted distributed ledger for robust authentication in healthcare cyber-physical systems. Sensors**24**(3), 938 (2024). https://doi.org/10.3390/s24030938
17. Rao, P.M., Deebak, B.D.: Lightweight two-factor authentication framework with privacy preserving for smart eHealth. Peer Peer Netw. Appl. **17**(1), 373–396 (2024). https://doi.org/10.1007/s12083-023-01596-5
18. Singh, B.M., Natarajan, J.: A novel secure authentication protocol for eHealth records in cloud with a new key generation method and minimized key exchange. J. King Saud Univ. Comput. Inf. Sci. **35**(7), 101629 (2023). https://doi.org/10.1016/j.jksuci.2023.101629
19. Saini, K.K., Kaur, D., Kumar, D., Kumar, B.: An efficient three-factor authentication protocol for wireless healthcare sensor networks. Multimed. Tools Appl. (2024). https://doi.org/10.1007/s11042-024-18114-1
20. Karimian, N., Saldamli, G., Park, Y., Lui, V.: Never lose your ECG: a novel key generation and authentication scheme for implantable medical devices. IEEE Access **11**, 81815–81827 (2023). https://doi.org/10.1109/ACCESS.2023.3302175

The manufacturer's authorised representative in the EU is Springer Nature Customer Service Centre GmbH, Europaplatz 3, 69115 Heidelberg, Germany. If you have any concerns regarding our products, please contact ProductSafety@springernature.com

Printed and bound by CPI Group (UK) Ltd, Croydon, CR0 4YY
16/09/2026
02217805-0001